biting at the PANORAMA, LEICESTER SQUARE.

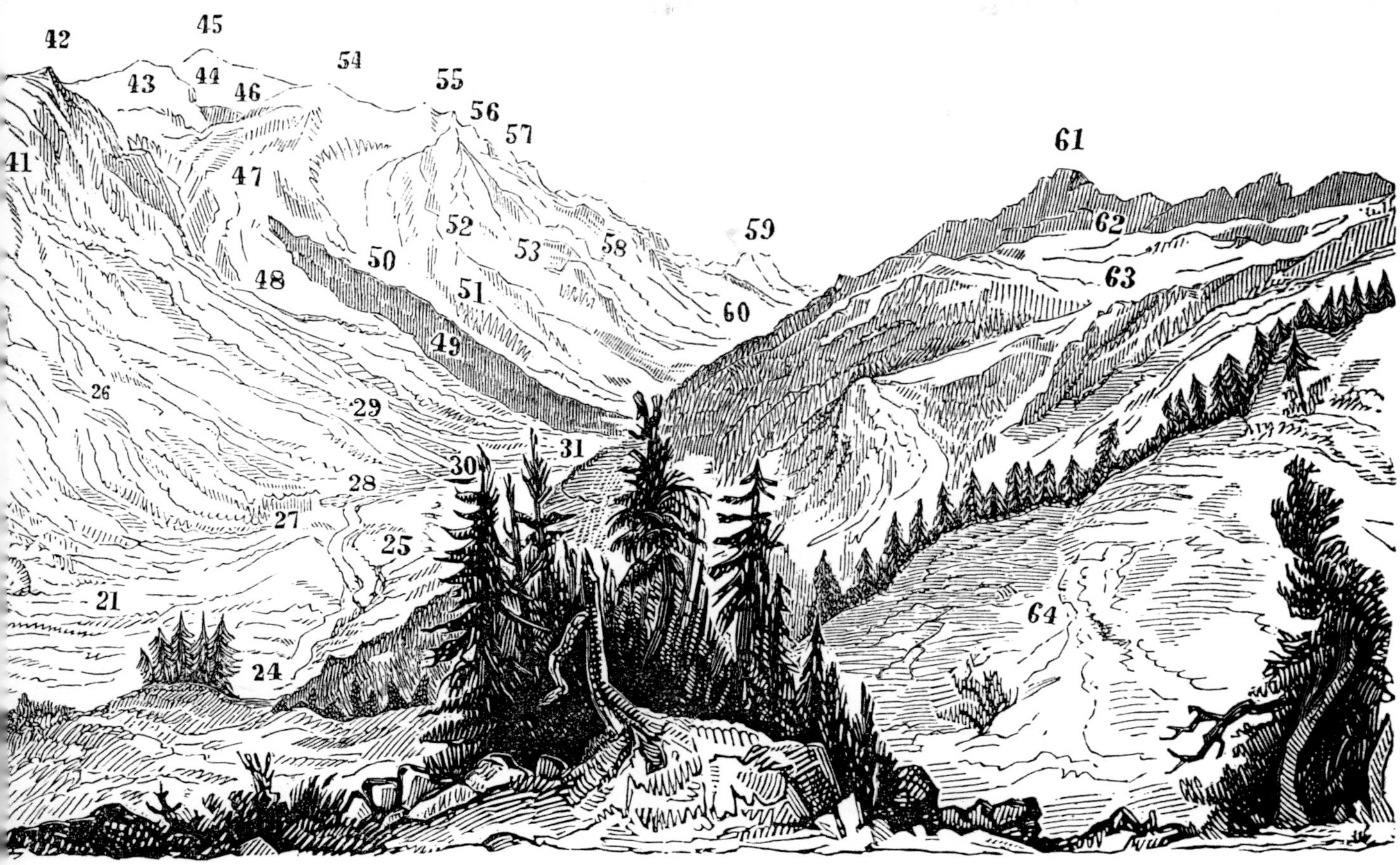

. Aiguille du Plan.
. Glacier de la Blaitière.
. Glacier du Plan.
40. Chalets des Blaitières.
41. Glacier des Pélerins.
42. Aiguille du Midi.
43. Mont Maudit.
44. Les grands Rochers Rouges.
45. Sommet du Mont Blanc.
46. Le grand Plateau.
47. Rochers des grands Mulets.
48. Glacier des Buissons.
49. Montagne de la Côté.
50. Glacier de Tacconay.
51. Montagne de Tacconay.
52. Glacier de Gria.
53. Montagne de Gria.
54. Dome de Goûte.
55. Aiguille de Goûte.
56. Aiguille du Bionnassay.
57. Aiguille du Miage.
58. Pierre Ronde.
59. Le Bonhomme.
60. Mont Lacha.
61. Le Breven.
62. Passage de la Cheminée.
63. Chalets de Plampra.
64. Sentier de la Croix de Flégère.

72. Hameau du Tour.
73. Glacier du Tour.
74. Aiguille du Tour.
75. Glacier d'Argentière.
76. Aiguille du Chardonnet.
77. Aiguille d'Argentière.

Flora of Chamonix

Flora of Chamonix

JOHN RUSKIN

EDITED AND INTRODUCED BY
DAVID S. INGRAM
AND STEPHEN WILDMAN

VOLUME TWO: COMMENTARY

PALLAS ATHENE

CONTENTS

Introduction
by Stephen Wildman and David S. Ingram
p. 7

RUSKIN AND CHAMONIX
by Stephen Wildman
p. 11

ALPINE BOTANY AND THE *FLORA OF CHAMOUNI 1844*
by David S. Ingram
p. 27

COLLECTING DAYS
with transcripts and notes by David S. Ingram
p. 55

RUSKIN AND BOTANY
by David S. Ingram
p. 133

Gazetteer of Chamonix and Environs, p. 150

Details of Illustrations, p. 152

Selected Bibliography and Further Reading, p. 154

Acknowledgements, p. 157

Index of Plant Species in Ruskin's *Flora of Chamouni 1844*, p. 158

General Index, p. 160

Opposite: George Richmond, John Ruskin (The Author of Modern Painters), 1843, photogravure after watercolour now destroyed. The distinctive profile of Mont Blanc appears in the background

Frontispiece (p. 2): John Ruskin, Aiguille Charmoz from window of the Hotel de l'Union, Chamonix, 1849, pencil, ink, and watercolour

Ruskin memorial stone, Chamonix,
incorporating medallion portrait by Michel de Tarnowski
and inaugurated 19 July 1925

Dedicated to the memory of the Ruskin scholar

JIM DEARDEN

who knew so much and helped so many

John Ruskin, Glacier des Bossons, Chamonix, *1849, pencil, ink, ink wash and bodycolour.*
A view looking down the valley of the Arve towards Les Houches, this is inscribed on the reverse: 'Glacier des Bossons from SE Window/Union'. Diary entries of 15 and 16 June 1849 record: 'Drew after dinner from window at Glacier des Bossons', and 'I have got on well with my drawing of the Glac[ier] des Bossons'.

INTRODUCTION

Published here for the first time is an intriguing example of the diversity of John Ruskin's interests. Geology and mineralogy had been his earliest youthful passions, his editor and biographer E. T. Cook declaring that 'No acquisition of later years – not his most radiant Turner or choicest manuscript – gave him pleasure so keen as he felt in the possession of his first box of minerals, and no subsequent possession, he says, had so much influence on his life.'[1] He formed significant collections of minerals and shells, much of which survive and have been the subject of study.[2] Recent research has also been undertaken on his interest in botany and its related published works such as *Proserpina* (1875-86), but one significant item has been hitherto overlooked – a collection of 46 pages of pressed flowers, with accompanying manuscript notes, bound in an album with the title *Flora of Chamouni 1844*, measuring 11 x 9¾ inches (28 x 25 cm). No mention of this can be found either in Ruskin's published writings or his manuscript diaries and letters, and it is not cited in the Library Edition (*The Works of John Ruskin*, ed. E. T. Cook and Alexander Wedderburn, 1903-12).

Its presence in the Whitehouse Collection – formerly held at Bembridge School, and since 1996 in the Ruskin Library, now re-named The Ruskin, at Lancaster University – suggests likely provenance from the Brantwood sale of 1931, at which John Howard Whitehouse, acolyte of Ruskin and founder of Bembridge School, acquired the bulk of his collection, but there is no evidence to confirm this. The album has rarely been mentioned in the Ruskin literature, although it has appeared in displays at the former Ruskin Library, notably *Ruskin's Flora* (2011).[3]

Such unpublished material is not quite unique. Two sets of botanical books, *British Phænogamous Botany* (W. Baxter, 2nd edition, 1834-43) and *English Botany* (J. E. Smith & J. Sowerby, 2nd edition, 1832-40), interleaved with copious annotations by Ruskin, surfaced only in 2015. Previously lost to sight, these have proved to be of considerable importance, described by David Ingram in a preliminary survey as 'a missing link in the chain of Ruskin's botanical studies.'[4] The *Flora* represents another link in that chain, although the combination of gathered flowers and notes speaks less of a focused botanical study and more of an additional record of an idyllic and fruitful holiday, in the company of his parents. Ruskin's commissioning of two professionally assembled *herbiers*, or albums of pressed flowers, from the young Venance Payot in Chamonix, is detailed in Stephen Wildman's chapter, 'Ruskin and Chamonix', while the

1. *The Works of John Ruskin*, ed. E. T. Cook and Alexander Wedderburn (George Allen, London, 39 vols., 1903-12), hereafter *Works*; vol. XXVI, *Deucalion*, p. xix.
2. Janet Barnes, 'The Mineral Collection of John Ruskin in the Ruskin Gallery, Sheffield', *UK Journal of Mines & Minerals*, no. 11, Spring/Summer 1992, pp. 26-29; Peter S. Dance, 'Ruskin the reluctant conchologist', *Journal of the History of Collections*, vol. 16, no. 1, 2004, pp. 35-46.
3. David [S.] Ingram and Stephen Wildman, *Ruskin's Flora: The Botanical Drawings of John Ruskin* (Ruskin Library and Research Centre, Lancaster University, 2011).
4. David [S.] Ingram, *Ruskin's Botanical Books: Re-ordered and Annotated Editions of Baxter and Sowerby* (The Guild of St George, 2016).

Carle Vernet, Les Voyageurs Anglais, c. 1830, lithograph
A gently mocking depiction of post-Napoleonic tourism, showing how the Ruskin family might have travelled

context and significance of the *Flora* are discussed by David Ingram in his 'Collecting Days' Editor's notes, and in 'Ruskin and Botany'.

John Ruskin took his final examinations at Oxford University in April 1842, receiving an honorary double fourth degree – by no means a disappointment, being the highest that could be awarded in the circumstances of having spent a year away from Oxford in 1840-41.[5] Following a three-month celebratory tour, focusing on Chamonix, he concentrated on writing the first volume of *Modern Painters*, whose publication in 1843 marked the beginning of his career as a critic. By 1844, travelling abroad again, the Ruskins (John and his parents, John James and Margaret) had already visited the Swiss Alps four times, sometimes accompanied by other family members.

5. Attested by Ruskin's biographer W. G. Collingwood, as cited in Tim Hilton, *The Early Years 1819-1859* (Yale University Press, 1985), p. 69.

That Chamonix was not actually in Switzerland was of little consequence: Ruskin's concern was for the physical geography (and geology) of place, without any restriction of national or political boundaries. His strictures on the simple innocence of Swiss pastoral life would implicitly include Chamonix and, later, Mornex near Geneva. What is now known as the Haute Savoie (a department in the Auvergne-Rhône-Alpes region) belonged historically to the House of Savoy, becoming part of the Kingdom of Sardinia in 1713 and then ceded to France in 1860. Ruskin considered this simply as 'the peaceful present of a few crags, goats, and goatherds by one king to another'[6], seeing it as potentially 'an immense benefit to Savoy ... A few millions of francs judiciously spent will gain to Savoy as many millions of acres of fruitfullest land and healthy air instead of miasma.'[7] He even stuck to the old name in his 1863 lecture, 'On the Forms of the Stratified Alps of Savoy'.

Stephen Wildman and David S. Ingram

Jean Dubois, Hôtel de L'Union, Chamonix, c. 1820, coloured aquatint

6. 'The Position of Denmark', *Morning Post*, 7 July 1864; *Works*, vol. XVIII, *Sesame and Lilies* etc., p. 549.

7. Letter to John Brown, 6 August 1860; *Works*, vol. XXXVI, *Letters*, p. 340.

John Ruskin, Mont Blanc and the Lake of Geneva from the Jura, 1835
pencil, ink and bodycolour

RUSKIN AND CHAMONIX

The Alps had exerted a fascination for the people of Northern Europe, not least the British[1], long before John James Ruskin and his wife Margaret took their fourteen-year old son John on a four-month tour of the continent in the summer of 1833, the first of some twenty-five tours in which Switzerland featured. After visiting the battlefield of Waterloo and travelling through Germany, they crossed the passes of the Splügen, Great St Bernard and Simplon before arriving at Chamonix (or in the older spelling Ruskin always used, Chamouni) on August 15th, affording John his first sight of a place which would become one of the fixed points in his life.[2]

'Before we left home,' he wrote in a fragmentary journal, 'I had read of Chamouni, heard of Chamouni, and seen some few drawings of Chamouni, but never so much as dreamed of going to Chamouni, it seemed so uncome-at-able; and for the Mont Blanc, it seemed in another world, in fairyland, and of course had a magic halo thrown round it, an ætherialness that can never be joined with reality.'[3] His cousin Mary Richardson, who was also in the party, recorded in her diary John's long walk with his father up to Montenvers and the source of the river Arveyron in the Mer de Glace. The visit gave rise to two of the best of Ruskin's early poems, 'The Arve at Chamouni' and 'Evening at Chamouni', as well as an essay published in *Loudon's Magazine of Natural History* for December 1834, 'Facts and considerations on the strata of Mont Blanc, and on some instances of twisted strata observable in Switzerland'. This confirms geology as his first passion, but beyond that was an immediate, deep and visceral attraction: 'There is not another scene like Chamouni throughout all Switerzland,' he wrote in his journal. 'In no other spot that I have seen is the rich luxuriance of the cultivated valley, the flashing splendour of the eternal snow, the impending magnificence of the bare, spiry crag, and the strange, cold rigidity of the surgy glaciers so dreadfully and beautifully combined. There is silence unbroken, no thunder of the avalanche comes crashing from the recesses of the hills, there is no voice from the chasmy glacier, no murmur from the thousand mountain streams, you are in solitude, a strange unearthly solitude, but you feel as if the air were full of spirits.'[4]

The next family tour lasted from June 2nd to December 10th, 1835, its itinerary helping to establish what would become Ruskin's 'old road', a familiar route through France and Switzerland into Italy. This time the goal was Venice, reached at the beginning of October after a stay of two months in the Alps. Ruskin had received as a fifteenth birthday present in 1834 a set of the *Voyages dans les Alpes* (1779–96) by Horace Bénédict de Saussure, whose interest in Alpine geology and glaciers is reflected in many of the entries in Ruskin's diary for 1835, often accompanied by careful pen and ink drawings.[5] The lyricism of 1833 is somewhat tempered, but on 8 July (the second in a stay of four days) he described the Col de Balme: 'As it thus commands the Bernese Alps on the one Side, and the Glaciers of Savoy on the other, there cannot be imagined a better observatory, or a

1. The poet Shelley visited Chamonix in 1816, acquiring 'a large collection of all the seeds of rare Alpine plants'; Charles I. Elton, *An Account of Shelley's Visits to France, Switzerland, and Savoy, in the Years 1814 and 1816* (Bliss, Sands, & Foster, London, 1894, p. 103).

2. For a detailed account of Ruskin's association with Chamonix, see André Hélard, *John Ruskin et les Cathédrales de la Terre* (Éditions Guérin, Chamonix, 2005).

3. *Works*, vol. II, *Poems*, p. 381.

4. *Ibid.*, p. 382. A digitisation project for *The Early Ruskin Manuscripts 1826–1842* is being undertaken by Prof. David Hanson of Southeastern Louisiana University: erm.selu.edu.

5. Whitehouse Collection, The Ruskin, Lancaster University, MS1.

J. M. W. Turner, Mer de Glace, from Liber Studiorum, *1812, etching and mezzotint*

more magnificent panorama to be observed.'[6] On page 31 of the manuscript is a tiny version of this, the peak of the Aiguille Verte 'covered with thick, white, and beautiful snow.'

To complement the diary Ruskin composed a long verse account of *A Tour through France*, culminating in the arrival at Chamonix:

By slow degrees and winding pathway mounting,
  Onward we went the ancient pines among;
By many a trickling stream and gushing fountain
  O'er the rough bed of many a torrent strong,
Amid the wreck of Servoz' ruined mountain,
  So beautiful in ruins. Far along,
And up the ridge, amid whose granites grey
  Struggles the Arve, you hear the constant cry
Of waters, though unseen they wind their way
  A hundred fathom down. In front, on high
The glacier flashes in the noontide ray,
  In the deep azure of the Alpine sky.
And now the ridge is won: rest, rest, and see
  What lies beneath you now; for there is Chamouni!
(Canto II, 23)[7]

Ruskin's undergraduate years at Christ Church, Oxford, were interrupted in 1840 by a bout of suspected tuberculosis, his terrified parents taking him on a nine-month tour of the continent to restore his health. This afforded barely a sight of the Alps, on the return in June 1841, but the family made up for it with a more focused tour in the summer of 1842, staying at Chamonix for a month from June 17th. '[W]e planned a summer-time in Switzerland,' Ruskin wrote in *Præterita*, 'not of travelling, but chiefly stay in Chamouni, to give me mountain air, and the long coveted power of examining the Mont Blanc rocks accurately. My mother loved Chamouni nearly as much as I; but this plan

6. Joan Evans and John Howard Whitehouse (eds.), *The Diaries of John Ruskin* (Clarendon Press, Oxford, 3 vols., 1956–59), hereafter *Diaries*; vol. I, p. 14. For an account of the tour, see Keith Hanley and Caroline S. Hull (eds.), *John Ruskin's Continental Tour 1835: The Written Records and Drawings* (Legenda, Cambridge, 2016).

7. *Works*, vol. II, *Poems*, p. 425; manuscript in the Beinecke Rare Book and Manuscript Library, Yale University.

John Ruskin, Aiguilles of Chamonix from Les Houches, 1842, pencil, watercolour and bodycolour

was of severe self-denial to my father, who did not like snow, nor wooden-walled rooms.'[8] The drawing above, described in a list of his manuscript diaries as 'Chamouni. First right study of rocks and foreground foliage', did indeed mark a maturity not only in careful observation but in the thought processes that would lead to the publication in 1843 of the first volume of *Modern Painters*, begun as a defence of his idol J. M. W. Turner. Apart from the St Gothard, Ruskin later noted that 'the valleys of Sallenches and Chamouni, and Lake of Geneva, are the only other Swiss scenes which seem to have made very profound impressions on him',[9] and he would have known the *Mer de Glace* subject in Turner's *Liber Studiorum* series of mezzotints, deriving from a visit to Chamonix in 1802. When writing about the year 1842 in his autobiography *Præterita* forty years later, Ruskin discovered the manuscript of his diary to be missing and had to admit that 'I find no diary whatever of the feelings or discoveries for this year. They were too many, and bewildering to be written. I did not even draw much—the things I saw were beyond drawing—but took to careful botany, while the month's time set apart for the rocks of Chamouni was spent in merely finding out what was to be done, and where.'[10] Very few drawings of 1842 are recorded, although one is a sizeable watercolour of *Mont Blanc with the Aiguilles, from above Les Tines*.[11]

On his return in August 1842 Ruskin wrote enthusiastic letters, the first to his college tutor the Rev. W. L. Brown: 'Chamouni is such a place! There is no sky like its sky. They may talk of Italy as they like. There is no blue of any firmament visible to mortal eye, comparable to the intensity and purity and depth of an Alpine heaven seen from 6000 feet up.'[12] And to his 'college friend' Edward Clayton he expressed a significant botanical interest: 'I got really rather fond of flowers at Chamonix; I intended to have sent you a drawing as you desired me, on the 8th, but

8. *Works*, vol. XXXV, *Præterita*, p. 312.
9. *Pre-Raphaelitism* (1851); *Works*, vol. XII, *Lectures on Architecture etc.*, p. 390.
10. *Works*, vol. XXXV, *Præterita*, p. 315.
11. Reproduced as the frontispiece to *Works*, vol. IV, *Modern Painters II*.
12. *Works*, vol. II, *Poems*, p. 223.

couldn't find one fit, for there nature uses them as I say—not to deck a bank, but to paint a mountain.'[13] Another poem followed, 'A Walk in Chamouni', published in *Friendship's Offering* for 1844, signed 'J. R., Christ Church, Oxford' and accompanied by an engraving from his watercolour of the Glacier des Bois.

Describing himself anonymously as 'a Graduate of Oxford' on the title page of the first volume of *Modern Painters*, published in May 1843, by the beginning of 1844 Ruskin was restless for the Alps. 'I am in one of those blue fits,' the diary notes on January 4th, 'in which one would be glad to throw up everything one possesses to get peace and live quietly in Chamounix. I fancy my summer's trip will put me right again.'[14] The family left Dover on May 14th, heading for the Alps with just that intention. 'We arrived at Geneva on 1st June, 1844,' he recalled in *Præterita*, 'with plan of another month at Chamouni; and fine things afterwards, which also came prosperously to pass. I had learned to draw now with great botanical precision; and could colour delicately, to a point of high finish. I was interested in everything, from clouds to lichens. Geneva was more wonderful to me, the Alps more living and mighty, than ever; Chamouni more peaceful.'[15] They stayed from June 6th to July 3rd, taking a tour to Zermatt before returning for another week, from the 24th to the 29th. A most propitious chance meeting took place in the Hôtel de la Poste at Simplon on July 16th, when Ruskin was 'forced' by his father to show his drawings to Professor James Forbes of Edinburgh University, winning the eminent geologist's 'sympathy and help ... his eye grew keen, and his face attentive, as he examined the drawings; and he turned instantly to me as to a recognized fellow-workman,—though yet young, no less faithful than himself.'[16] Not at that time familiar to him, Forbes's book of 1843, *Travels through the Alps of Savoy and other Parts of the Pennine Chain*, would become an invaluable aid.[17]

The 'collecting days' relevant to the compilation of the *Flora*, covering much of the stay, are described in detail in David Ingram's commentary below. Others were no less busy, with just a few more restful, towards the end: 1 July, according to the diary, was 'a delicious day: couched among Alpine roses, alone, not even Coutet [sic] with me; not a human sound for hours'.[18] Appearing in the diary simply as 'a new guide', Joseph-Marie Couttet 'came to us on the morning of the 7th of June. My father explained to him that he wanted me taken charge of on the hills, and not permitted in any ambitious attempts, or taken into any dangerous places; and that, from what he had heard of Couttet's trust-worthiness, and knowledge of his mountains, he had no doubt that I should be safe with him, and might learn more under his tutelage, in safety, than by the most daring expeditions under inferior masters. Couttet said little, but accepted the charge with a kindly glitter in his eyes, and a cheerful word or two, signifying that my father need not fear for me; and we set out together for the base of the Buet,—I on muleback, he walking. For thirty years he remained my tutor and companion. Had he been my drawing-master also, it would have been better for me: if my work pleased Couttet, I found afterwards it was always good; and he knew perfectly when I was trying vainly to do what I could not, or foolishly what no one else would care for. The month at Chamouni, however, passed with his approval, and to my perfect benefit. I made two foreground studies in colour, of considerable beauty; and, under his teaching, began to use my alpenstock easily, and to walk with firmness.'[19] Couttet, born in 1792, was only seven years younger than John James Ruskin. The son of a guide who had acted for Saussure, he was called by Ruskin 'the captain of Mont Blanc', having climbed it a dozen times, and came to hold a place in Ruskin's affections which would last for thirty years.

The remarkably fruitful stay of 1844 had an interesting sequel, until now wholly overlooked. Needing to study early Renaissance painting in Florence and Venice, in 1845 Ruskin travelled abroad without his parents for the first time, although he did take something valuable from Chamonix with him. 'My father could not spare time to go with me,' Ruskin recalled, 'so

13. *Works*, vol. I, *Early Prose Writings*, p. 474.
14. *Diaries*, vol. I, p. 258.
15. *Works*, vol. XXXV, *Præterita*, p. 328.
16. The meeting is recorded in the diary entry of July 16th, 1844 (*Diaries*, vol. I, p. 302) and described in *Deucalion*, vol. I, ch. X (1876); *Works*, vol. XXVI, *Deucalion*, pp. 219–20.
17. A copy of the second edition (1845) is recorded in Ruskin's library, and was doubtless in use on later visits.
18. *Diaries*, vol. I, p. 295.
19. *Works*, vol. XXXV, *Præterita*, p. 329.

Karl-Rudolph Weibel-Comtesse, Joseph Couttet, aged 55, 1847, watercolour and bodycolour

he asked me to take my Chamouni guide, Joseph Couttet, by way of pro-papa. He was a tutor, and domestic Pope's legate, of perfect fidelity and good sense: a good practical physician also; I never had occasion to call in any other; and he always after that time travelled with me when my father and mother could not, (my mother never left my father,) until Couttet's death in 1875. He was nearly fifty when, in 1845, he met me at Geneva in early April [in fact Couttet was 52]; and we travelled leisurely through Lower Savoy and Provence to Fréjus. It was starlight, after a long day's drive, as we came down towards the sea over the southern moors of wild myrtle; and I recollect teasing old Joseph considerably by humming 'com' è gentil' all the way.'[20]

Having spent nearly six weeks in Florence, and realising how much time he would need in Venice, Ruskin told his father on July 27th that he would 'bolt over Simplon or Cenis home—merely calling at Chamonix to drop Coutet [sic] & pick up a flower book which Payot is making for William, and another of the rarest plants for myself, on an extensive scale.'[21] By August 17th, still two weeks before reaching Venice, he was already revising his plans: 'I shall not I think even call at Chamouni. I should not like to go there merely to go away again—shall make them bring my books to Martigny to be ready as I pass, and so straight by Vevay over Jura, not even touching Geneva.'[22] A later letter, written from Noyon on October 25th, told his father that he had indeed been 'obliged to let Coutet [sic] go at Martigny ... I have promised him nothing. He left me at six o'clock this morning. His address is simply Chamonix, but I suppose his money must be sent to Geneva.' Often neglectful over paying those who had done him loyal service, Ruskin was expecting

20. *Works*, vol. IV, *Modern Painters II*, p. 345.
21. Harold I. Shapiro (ed.), *Ruskin in Italy: Letters to his Parents 1845* (Clarendon Press, Oxford, 1972), p. 162.
22. *Ibid.*, p. 177.

his father to meet a bill not only for Couttet's attendance, 'April 12th to Oct 26th, 198 days, call it 200 ... 800 Fr[ancs]', but also for 'Plants bought from Payot—80 [Francs].'[23] If the *Flora of Chamouni 1844* were not known to exist, these references would be mysterious. Clearly, Ruskin had ordered two 'Floras', albums of pressed flowers (in French, *herbiers*), probably through Couttet, perhaps even having instigated the commission before leaving Chamonix in 1844; one was presumably intended as a gift for his cousin William Richardson, brother of Mary. Venance Payot, who was only eighteen years old in 1844, was the son of a mountain guide, accompanying his father Pierre Joseph on a trip to the Mer de Glace with Alexandre Dumas. Developing into a respected naturalist and collector of plants, he is known to have assembled albums superficially similar to the *Flora of Chamouni 1844*.[24] These are less impressively bound than Ruskin's, while scientifically they are more complete (see David S. Ingram, 'Alpine Botany and the *Flora of Chamouni 1844*'), and although it seems that he must have supplied Ruskin with *herbiers*, the *Flora* itself is clearly not his work. No further information can be found.

Study in Italy enabled Ruskin to complete the second volume of *Modern Painters* (published in April 1846), and as he put it in *Præterita*, 'As I had done my second volume greatly to my father's and mother's delight, (they used both to cry a little, at least my father generally did, over the pretty passages, when I read them after breakfast) it had been agreed that they should both go with me that summer to see all the things and pictures spoken of,—Ilaria, and the Campo Santo, and St. Mary's of the Thorn, and the School of St. Roch.'[25] By this he meant Lucca, Pisa and Venice (as well as Florence), but there was time for four days in August to visit Chamonix, near the end of a six-month tour, where, 'having seen enough by this time of the upper snow, I was content to enjoy my morning walks in the valley with papa and mamma; after which, I had more than enough to do among the lower rocks and woods till dinner time, and in watching phases of sunset afterwards from beneath the slopes of the Breven.'[26] The sole Chamonix diary entry (August 23rd) includes another striking word picture: 'Went out again in the afternoon towards the Cascade des Pélerins. Surprised to see the real rain-clouds assume on the Breven, about one third of its height, the forms of cirri,—long, continuous, and delicate; the same tendency showing in the clouds all along the valley, some inclining to the fish shape, and others to the cobweb-like wavy film.'[27]

He also had the company of his Oxford friend Henry Acland, a medical man who shared an interest in geology, as Ruskin wrote in a letter a week later (August 30th) to the artist George Richmond: 'I was only there [Chamonix] four days myself. I didn't want to go at first, because it always gives me too much vexation to leave it. But we went because it was said some rocks were bared on the Mont Blanc in unusual places. All newspaper [i.e. false reporting]—the Mont Blanc is as changeless as the blue sky above it; but though we had wretched weather, I never thought Chamouni so unearthly—it is quite awful, and quite alone—nothing that I have yet seen can be compared with it in any wise; its inexhaustibleness and perpetual freshness to me I am truly thankful for—other scenery palls. I never entered it with so much wonder, nor left it with so strong regret; when you come abroad you should really go there, and not to Italy.'[28]

This is a good point to consider what Chamonix had already come to mean to Ruskin, and how much more it would enter his soul. Even at the age of twenty-two, arriving in Venice in 1841 he could confide to his diary that 'the Paradise of cities' and Chamonix 'are my two bournes of earth'.[29] By 1844 it had become 'my own valley'; in 1853 he told his father that 'there is nothing like it', and in 1854 he noted that 'Every day here I seem to see further into nature, and into myself—and into futurity.'[30] To his friend Jemima Blackburn, who had asked him in 1855 to name his favourite place, Verona rather than Venice had become

23. *Ibid.*, p. 231.
24. Venance Payot, *Herbier des Alpes* (Les Éditions de l'Amateur, Paris, 2006) reproduces specimens from a private collection. For Payot, see https://www.lectura.plus/expositions/payot/
25. *Works*, vol. XXXV, *Præterita*, p. 415.
26. *Ibid.*, p. 420.
27. *Diaries*, vol. I, p. 344.
28. Letter to George Richmond, 30 August 1846; *Works*, vol. XXXVI, *Letters*, p. 63.
29. *Diaries*, vol. I, p. 183 (May 6th, 1841).
30. *Diaries*, vol. II, p. 498 (July 10th, 1854).

John Ruskin, Aiguilles of Chamonix from the foot of La Flégère, 1849, pencil and watercolour

the alternative choice: 'If, for an affection, you want a *narrower* answer than Chamouni, I am a little puzzled between the top of the Montanvert and a small rock on the flank of the Breven. I have been *happiest* on the Montanvert [the closest point for access to the Mer de Glace glacier], but *oftenest* at this rock, where I generally pass my evenings when at Chamouni [the rock on the Brévent is on the other side of the valley, looking over Chamonix towards Mont Blanc]. Next to the valley of Chamouni, and even running it rather hard, I love the little Scaliger churchyard at Verona. I think I have been more intensely happy for a little while in the churchyard, but not so enduringly.'[31]

In the 1854 diary Ruskin compiled a detailed 'Catalogue of Sketches in neighbourhood of Mont Blanc', 64 altogether and dated 1844 or 1849 (with the exception of the 1842 *Aiguilles from Les Ouches*); the known watercolours so far located all carry a C mark and catalogue number.[32] Switzerland was the principal focus in 1849 (when he was again with his parents), his diary described as '*Chamouni and Simplon. Glacier work begun.*', based on a month in Chamonix from June 13th. After tea on the 18th he sought his 'beloved old place on the Breven and saw a solemn sunset … Thank God for permitting me to sit on that slope once more, thus strong in health and limb.'[33] One of the finest of all his

31. *Works*, vol. XXXVI, *Letters*, p. 193.
32. Listed in *Works*, vol. V, *Modern Painters III*, pp. xxi–xxii.
33. *Diaries*, vol. II, p. 391.

watercolours, the *Cascade de la Folie*, is identifiable as a view from the Hotel de l'Union. Drawing exactly what he saw on the spot was a fundamental principle, so such a highly finished work could only have been made from a fixed and favoured place. According to contemporary guide books the Union was not quite as smart as the Hôtel de Londres opposite, later incorporated with the Hôtel de l'Angleterre, but this choice was typical of the Ruskins on their European travels (as in the nine-month tour of 1840–41), always preferring the slightly less ostentatious, and usually more comfortable and accommodating hotels.[34] They would have been a little cheaper, too, suiting John James Ruskin's thrifty nature. An added benefit of the Ruskins' love of the 'old road' was that the proprietor families of the Union—Charlet, Simond and later Eisenkrämer—could be counted almost as extended family: in 1874 Ruskin mused that '[e]ven in these days, I have had no more true or kind friend than the now dead Mrs. Eisenkraemer of the old Union Inn at Chamouni; and an innkeeper's daughter in the Oberland taught me that it was still possible for a Swiss girl to be refined, imaginative, and pure-hearted, though she waited on her father's guests, and though these guests were often vulgar and insolent English travellers.'[35]

While his parents returned to Geneva in July 1849, John was allowed further close study of the Alps, describing himself being 'as busy as an ant at Chamonix', also reaching Zermatt and the Matterhorn in the company of his manservant John Hobbs and Couttet. A highlight was a three-night stay in the rustic travellers' refuge on the Montenvers, which left him in near ecstasy: 'I think I never enjoyed any evening so much as this in my life, unless it were one at Champagnole in 1845. I had no idea what this place was, until I sat at the window quietly to-day watching the sunset and the vast flow of the ice, welling down the gorge—a dark and billowy river—yet with the mountainous swell and lifted crests that the iron rocks have round it. I have been nearly all day drawing at the Aiguille Blaitière.'[36] This was written on August 22nd; three days later he could record that 'I have certainly not lately, nor often in old times, felt stronger emotion than in watching the dawn from the Montanvert these three mornings past. Yesterday I saw it when it was still very dark, and Orion burning beyond the Grandes Jorasses; and the whole river of heaven between the hills full of stars; and again, later, when I was watching the increase of the serene clear cold morning light, a beacon intensely bright flashed out on the summit of the Dru; it was the morning star.'[37] Only a week earlier he had been moved almost beyond words: 'I never saw the valley look so lovely as it did to-night, with its noble quiet slopes of deep, deep green and grey; and above them the rich orange of the Aiguilles ... I was almost in tears as I watched the light declining behind the grand pines' sweep and rugged crest of the noble Breven once more.'[38]

Such purple prose impressions complement diary entries full of detailed observation of nature, including clouds, rocks and flowers: 'The Alpine roses are just bursting at the Montanvert, but I will not describe it in the rain, though I never thought it so lovely.'[39] By this date Ruskin had acquired his own daguerreotype equipment, several of the 1849 watercolours matching subjects taken by the indefatigable Hobbs.[40] A diary entry for June 28th, of scrambling near the Cascade des Pèlerins, tells that he 'got some pines, in Dag[uerreotype], then finished my drawing successfully.'[41] Hobbs's own fragmentary diary tells that on June 22nd-23rd he '[g]ot a beautiful Dag. of a piece of glacier', possibly the *Mer de Glace* plate echoing exactly his master's drawing. Poor

34. 'Messrs. Charlet and Simond took the Hotel de l'Union at the commencement of the season of 1838, and have done themselves much credit by the excellent and liberal manner they have conducted it.' *The European Indicator or Road-Book for Travellers on the Continent* (Imprimerie Félix Le Monnier, Florence, 1841), p. 70. Information on the Chamonix hotels is given in W. A. B. Coolidge, *Swiss Travel and Swiss Guide-Books* (Longmans, Green, and Co., London, 1889).
35. *Fors Clavigera*, Letter 44, August 1874; *Works*, vol. XXVIII, *Fors Clavigera*, p. 131.
36. *Diaries*, vol. II, p. 431.
37. *Ibid.*, p. 432.
38. *Ibid.*, p. 425 (August 15th).
39. *Ibid.*, p. 386 (June 14th).
40. See Ken and Jenny Jacobson, *Carrying Off the Palaces: John Ruskin's Lost Daguerreotypes* (Quaritch, London, 2015).
41. *Diaries*, vol. II, p. 397.

Opposite: John Ruskin, Cascade de la Folie, Chamonix, 1849, pencil, watercolour and bodycolour

John Ruskin and John Hobbs, Mer de Glace, Chamonix, 1849, daguerreotype. Possibly the image recorded by Hobbs in his diary for June 22nd-23rd, 1849 as 'a beautiful Dag[uerreotype] of a piece of glacier'

Hobbs, trying to keep up with Ruskin and Couttet in search of a good view of the Aiguilles du Midi—'as I had my arms full with the Daguerreotype machinery it was no joke I assure you, jumping from stone to stone'.[42]

On both outward journeys of their long stays in Venice over 1849–50 and 1851–52, Ruskin and his new wife Effie spent several days in Chamonix. In 1849 this was between October 14th and 17th. 'It is most amusing,' Effie wrote to her brother George, 'to see the surprise and delight of the people at the return of John. They seem fond of him and they pay me pretty little compliments.'[43] Two years later they were there on August 13th to 15th, still loyal to the Hôtel de l'Union. Accompanied by Charles Newton and the Rev. Daniel Moore of the Camden Chapel, Camberwell, and in spite of poor weather, Ruskin managed some of his favourite long walks, to the source of the Arveyron, Montenvers and the Glacier des Bossons. Effie also made it by mule up to the Mer de Glace, spending the night with John in the travellers' refuge; 'it is a splendid day' she noted, '& I have been sitting outside the house working on the bank surrounded by goats and watching the people coming up in such quantities as makes it quite an Exhibition of the Nations.'[44] 'I love the place better than ever,' Ruskin told his father, 'and think it lovelier, and I don't know that I was ever sorrier to leave it.'[45]

Sadly, by the time of his next visit in the summer of 1854 the ill-fated and famously unconsummated marriage had ended in annulment, after Effie left Ruskin for the painter John Everett Millais (whom she married in that year). A return to Chamonix was therefore hugely welcome, as recorded in his diary for July

42. John Hobbs, Manuscript Diary, Morgan Library, New York (MA 2539). Charmingly, stuck to its last page is a pressed flower, with the inscription 'Gentian from the Alps at Chamouni / collected at this date / August 1849'; illustrated in David Hill, 'Ruskin Drawings at King's College, Cambridge: #4 The Mer de Glace from the Montanvers Hotel above Chamonix', *Sublime Sites*, https://sublimesites.co/2018/12/06/ruskin-drawings-at-kings-college-cambridge-4-the-mer-de-glace-from-the-montanvers-hotel-above-chamonix/

43. Mary Lutyens (ed.), *Effie in Venice* (Pallas Athene, London, 2003), p. 47.

44. *Ibid.*, p. 181.

45. *Works*, vol. XXXVI, *Letters*, p. 117 (August 16th, 1851).

John Ruskin, Mer de Glace
from the Montanvers Hotel above Chamonix, 1849,
pencil, ink and watercolour

10th, the day he arrived: 'Thank God, here once more, and feeling it more deeply than ever. I have been up to my stone upon the Breven, all unchanged and happy.'[46] A fortnight's stay included both introspection and careful, if sometimes romantic, botanising: 'Note that the wild roses, here in Chamouni, have the most perfect perfume I ever felt, like sweet briar' … 'Yesterday up at the old place under Aiguille Blaitière and on the Charmoz; the hill up to the glacier covered with the pink moss 'Cyllene Acaulis'? and a crested purple flower 'Pedicularis Rostrata'? mixed with small but exquisitely coloured forget-me-not and white saxifrage'.[47] Returning a month later, more time was spent on geology than botany. On these visits in 1854, and again in the summer of 1856 (between July 26th and August 18th), Ruskin set his manservant Frederick Crawley (successor to Hobbs from late in 1852) to make daguerreotypes of subjects in Chamonix—as many as sixteen in 1854—and in Basel, Thun and Fribourg, more than twenty probably made in 1856.[48] Two of the most beautiful and dramatic, *Aiguilles, Chamonix* and *Aiguille Verte and Aiguille du Dru, Chamonix*, are dated on the mounts by Crawley to 1854.

Just as earlier visits had contributed to themes within the first two volumes of *Modern Painters*, so the extended stays of 1849 and 1854 bore fruit in the next two, published together in 1856. In Chapter III of Volume III, 'Of Turnerian Light', Ruskin uses an experience at Chamonix, 'on a warm sunny morning, about eleven o'clock, the sky clear blue', to demonstrate the extreme difficulty—although Turner comes close—of capturing on paper or canvas what the eye can see: 'it will thus generally be found impossible to represent, in any of its *true* colours, scenery distant more than two or three miles, in full daylight. The deepest shadows are whiter than white paper.'[49] No other writer on art, or landscape, had ever made such heartfelt empirical observation. Chamonix necessarily figures in Chapter XIII, 'Of the Sculpture of Mountains:—Secondly, the Central Peaks', as part of the system of rock whose 'jagged and spiry outlines' lead the eye towards Mont Blanc. Ruskin balances geology against divine

46. *Diaries*, vol. II, p. 498.
47. *Ibid.*, p. 499 (July 20th and 22nd).
48. Jacobson, *Carrying Off the Palaces*, 2015, cat. nos. 234–246, 264–306.
49. *Works*, vol. V, *Modern Painters III*, pp. 54–55.

John Ruskin, Aiguille Structure, Modern Painters, *Vol. IV, 1856, engraved by J. C. Armytage: 'jagged and spiry peaks', seen above 'two long slopes of comparatively flattish mountain'*

creation, appreciating that both tell a history of 'endurance and destruction'. He was among the first to realise, not least from personal experience over several decades, that glaciers such as the Bossons and even the Mer de Glace showed signs of receding, as part of an inevitable natural process.[50]

Less palatable was a sense of change for the worse as a result of human activity. Youthful admiration for an idealised view of simple Alpine life gave way to feelings of disillusion, casting a shadow in Chapter XX, 'The Mountain Glory': 'I believe that every franc now spent by travellers among the Alps tends more or less to the undermining of whatever special greatness there is in the Swiss character. … The influx of foreigners into Switzerland [in which Ruskin included Chamonix] must necessarily be greater every year, and the greater it is, the larger in the crowd will be the majority of persons whose objects in travelling will be, first, to get as fast as possible from place to place, and, secondly, at every place where they arrive, to obtain the kind of accommodation and amusement to which they are accustomed in Paris, London, Brighton, or Baden.' In a following passage of pessimistic prophecy, he even saw his favourite place on earth to be under threat: 'The valley of Chamouni, another spot also unique in its way, is rapidly being turned into a kind of Cremorne Gardens; and I can foresee, within the perspective of but few years, the town of Lucerne consisting of a row of symmetrical hotels round the foot of the lake, its old bridges destroyed, an iron one built over the Reuss, and an acacia promenade carried along the lake-shore, with a German band playing under a Chinese temple at the end of it, and the enlightened travellers, representatives of European civilization, performing before the Alps, in each afternoon summer sunlight, in their modern manner, the Dance of Death.'[51] Ironically, there is evidence that Ruskin himself had contributed to this fashion: in an account of one of Thomas Cook's earliest conducted tours to Switzerland, a passage from 'The Mountain Gloom', Chapter XIX of Volume IV, was quoted by its correspondent, commissioned by the Alpine Club in 1863. Her botanical interest, however, ran only to gathering blooms from the Alpine rhododendron to 'deck our hats with its branching sprays, and pluck bouquets from either side of our path which here and there is literally ribboned with a border of variegated flowers.'[52]

50. 'The Glacier de Bois is no more,' he told Charles Eliot Norton in 1874; J. L. Bradley and I. Ousby (eds.), *The Correspondence of John Ruskin and Charles Eliot Norton* (Cambridge University Press, 1987), p. 343.

51. *Works*, vol. V, *Modern Painters III*, pp. 455–56.

52. *Miss Jemima's Swiss Journal* (Putnam, London, 1963), p. 37; the author's surname has not been identified. This passage was included in the first volume of *Selections from the Writings of John Ruskin* (Smith, Elder, and Co., London, 1861).

In July 1856 Ruskin was in Chamonix with his parents, on a visit to Switzerland which also saw the development of a new friendship with the American scholar Charles Eliot Norton, one which would remain lifelong, the future Harvard Professor eventually acting as one of Ruskin's literary trustees. Complementing the taking of daguerreotypes by Crawley was some serious geological study: the morning of July 29th saw him '[at] work with pickaxe and spade before breakfast for an hour and a half (out at six). Left man at work to continue.'[53] Not wanting to neglect his parents, he went with them the next day 'to old rock where we used to sit, in the Breven wood. ... Read *La dame aux cheveux gris* all the evening to my mother.'[54] Two years later, on the return leg of a tour to Rheinfelden, Bellinzona and the southern Alps following in Turner's footsteps, he took time to make an even more thoughtful if fleeting visit. It was just a single day, but as he told his father in a letter of September 8th, 1858, 'I have enjoyed my day much and *felt* Chamouni as much as ever.' The locals, including the Eisenkrämers, were 'all glad to see me, & I think they were much gratified at my coming with the sole purpose of paying them a visit.'[55]

The Union Hotel was again Ruskin's destination in the summer of 1860. As in 1846 and 1856, his reward for completing volumes of *Modern Painters*—Volume V was issued on June 14th, 1860—was a stay in the Alps. Leaving England on May 22nd, he reached Geneva on the 28th, and later recalled that he received the book at St. Martin's, a regular stop on the old road, 'where I was then resting, chiefly for the sake of the gentians, lilies, and wild roses of the Aiguille de Varens. ... I went up thence to Chamouni, and there wrote *Unto This Last*, reading it, as it was finished, to the friend staying at the Old Union Inn with me, Mr. Stillman.'[56] The transition in Ruskin's writings from art history to the moral and social issues increasingly addressed in the latter stages of *Modern Painters* has been commonly observed, and although weary, he was able, using as sounding-post another American friend, William James Stillman, to develop critical economic ideas in what would become a bombshell of a book. *Unto This Last* was first published as four essays in *The Cornhill Magazine* between August and November 1860 (and in America, in *Harper's New Monthly Magazine* between September and December), then in book form in 1862, remaining perhaps the most widely read of all his works, still regarded as influential at the end of the century. Sadly, neither diary nor letters from this most exceptionally fruitful time in Chamonix have survived.

Ruskin must have expected adverse reaction to his bold critique of capitalism, but was surprised by the level of hostility to what he called his 'little book', and was undoubtedly discouraged, to the extent of considering permanent life abroad, ideally in Switzerland. Even though he had told his friend Dr John Brown in August 1860 that he thought 'Chamouni itself and all the rest of Switzerland ... completely spoiled by railroads, huge hotels, and architects out of employ',[57] he spent two months in Lucerne in the autumn of 1861 and went ahead with a move to Mornex, near Geneva, in August 1862, renting what is now known as the Pavillon des Glycines (formerly occupied by Richard Wagner) until May 1863, bar a month back in London in November. Despite having a bedroom window view of Mont Blanc, Ruskin was unhappy with its proximity to the city and looked for the opportunity to buy land, both at Brison near Bonneville (which he called the Brezon, using an older name), and at Chamonix. An excited letter to John James Ruskin just before leaving reported the purchase, through Eisenkrämer, of 'the very piece I always was so fond of, with the two châlets under the Aiguille Blaitière.'[58] 'I bought just before leaving Savoy about a square mile of Mont Blanc—2000 feet above Chamouni with quantities

53. *Diaries*, vol. II, p. 518.

54. *Loc. cit.* A novel of 1854 by Henriette Cabrières (Madame Geisendorf).

55. John Hayman (ed.), *John Ruskin Letters from the Continent 1858* (University of Toronto Press, Toronto, 1982), p. 168.

56. *Works*, vol. XIII, *Turner*, p. 497.

57. *Works*, vol. XXXVI, *Letters*, p. 340 (August 6th, 1860).

58. *Ibid.*, p. 445 (May 26th, 1863). 'I first actually bought the piece of meadow in Chamouni above the chalets of Blaitière; but sold it on perceiving what ruin was inevitable in the valley after it became a tourist rendezvous.' (*Præterita*, vol. II, Ch. 11; *Works*, vol. XXXV, p. 436). This must have been after 1869, as he reported to his friend Lady Mount-Temple that 'I got a letter only the day before yesterday from Chamouni, saying the people still believed a certain plot of hillside to be mine.' (John Lewis Bradley (ed.), *The Letters of John Ruskin to Lord and Lady Mount-Temple*, (Ohio State University Press, Columbus, 1964), p. 202, letter of May 22nd, 1869).

of weeds and stones upon it—and a torrent—and I want to buy a glacier above', he told Pauline Trevelyan.[59] As much rock and wood as useful ground—just as Ruskin's later acquisition at Brantwood could perhaps be described—it appealed to him not least æsthetically, as he had drawn it more than once: 'The principal smooth bit of it,' he informed his father, 'is that on right hand in the finished grey sketch of Chamouni in your room.' Ruskin returned in the autumn, now reporting Couttet's help in identifying title to the land, but with certain complications. John James, asked to send credit of £1000 (£720 the land, plus £200 for a bemused George Allen, brought to Switzerland as Ruskin's assistant and somewhat stranded in Mornex), must also have been concerned to read expressions of misgiving, probably explained more fully on his son's return in November. 'The only thing that grieves me,' Ruskin had admitted, 'is when these old mountain feelings pass from me',[60] as they seemed now to do. The dream had passed too, and he would not see Chamonix again for more than a decade.

Future visits to Switzerland between 1866 and 1872, more akin to holidays in various company, concentrated on the quieter and less troublesome tourist destination of Giessbach, on Lake Brienz. Ruskin's disappointment in the failure of an impossible idyll would have been lessened by an increasing sense of change in the places he loved most, none of it for the better. As he wrote to Norton in January 1866: '*Every* thirteenth-century cathedral in France, and every beautiful street in my favourite cities, has been destroyed. Chamouni is destroyed—Geneva—Lucerne—Zurich—Schaffhausen—Berne,—might just as well have been swallowed up by earthquakes as be what they are now. There are no inns,—no human beings any more anywhere—nothing but endless galleries of rooms, and Automata in millions.—I can't travel—I have taken to stones and plants.'[61] When he did finally visit Chamonix again, this feeling was mixed with at least some revival of past happiness and fond memories. Having spent the spring and summer of 1874 in Italy, including a trip to Sicily, Ruskin made time for two excursions from Geneva on the way home in October, managing what proved to be a last meeting with his old friends the Couttets. Writing to Mrs John Simon, he bemoaned the appearance of large houses and a 'huge monster of an inn by the church'[62], while still finding half the Montenvers path, at least, as he had first known it. To his cousin Joan Severn he expressed delight that 'there is actually not a rock, not a cottage, not a tree, traceably changed in forty years. I am writing in a room which seems to me the same in wall, floor, furniture, that it was when my father and mother and I first saw Mont Blanc near, from it, in 1833. Nothing is changed—except Mont Blanc and I! Both of us fatally. He, as thin and wasted in snow, as I in heart and flesh. I never saw anything so sorrowful, so unbelievable, in the courses of nature. This great mountain, which one thought so eternal, faded like a white rose.' But there was one significant change, in an aspect of modern life constantly on his mind: 'The road, which once was little more than a beautiful winding lane, well made, is now everywhere widened and in most places levelled and straitened. This enables vast omnibuses, like our tramway ones, to carry the mob up to Chamouni from Geneva in a day. Away they bowl, as if to Epsom; the harder they go the merrier. And I don't believe one fool in a dozen dozen [sic] so much as looks up at the cliffs on each side, or at a flower, or a tree, in the most wonderful piece of mountain scenery I know in the world.'[63]

Fourteen years passed before Ruskin's last visit in 1888. In the company of Arthur Severn, it would prove to be his final sight of the continent, starting in Abbeville and Paris en route to an unhappy valediction in Venice, where his health broke down. 'I have to see my old Chamouni on the way,' he told Francesca Alexander, writing from Paris,[64] and the week in Chamonix in mid-September proved to be a high point.[65] Here he achieved one of the summer's principal purposes, to write an Epilogue to a new

59. Virginia Surtees (ed.), *Reflections of a Friendship: John Ruskin's Letters to Pauline Trevelyan 1848–1866* (George Allen & Unwin Ltd., London, 1979), p. 214 (letter of June 17th, 1863).
60. *Works*, vol. XXXVI, *Letters*, p. 454 (September 14th, 1863).
61. Bradley and Ousby, *The Correspondence of John Ruskin and Charles Eliot Norton* (1987), p. 95 (January 13th, 1866).
62. *Works*, vol. XXXVII, *Letters*, p. 146 (October 15th, 1874).
63. *Works*, vol. XXIII, *Val d'Arno*, p. li (October 6th, 1874).
64. Lucia Gray Swett, *John Ruskin's Letters to Francesca and Memoirs of the Alexanders* (Lothrop, Lee & Shepard Co., Boston, 1931), p. 175 (letter of August 25th, 1888).
65. The penultimate chapter of *Præterita*, 'L'Esterelle', was written on

John Ruskin, Mer de Glace, Chamonix, 1874, watercolour.
This drawing probably relates to the diary entry of October 15th, 1874: 'Up to Montanvert, and fresh and strong in work'

Complete Edition of *Modern Painters*, the first to appear wholly under George Allen's imprint. '[I] have just written the last clause of the epilogue to *Mod[er]n Painters*,' he confided to the diary on Sunday September 16th, 'in the perfectest light of Mont Blanc, after being at mass, and a little walk on fresh grass towards source of Arveron.'[66] In what amounts to his last significant piece of writing, nine paragraphs offer a kind of *nunc dimittis* to a life's work. In *Unto This Last*, mostly written at Chamonix nearly thirty years earlier, Ruskin had concluded that 'There is no wealth but life.' Now he could remind his readers of two other fundamental conclusions—that 'All great Art is Praise' (first voiced in *The Laws of Fésole* in 1877), and from the 1870 *Lectures on Art* that 'Life without Industry is sin, and Industry without Art, brutality.'[67] The final sentence is an apt summation of fifty years' inspiration: 'and now, in writing beneath the cloudless peace of the snows of Chamouni, what must be the really final words of the book which their beauty inspired and their strength guided, I am able, with yet happier and calmer heart than ever heretofore, to enforce its simplest assurance of Faith, that the knowledge of what is beautiful leads on, and is the first step, to the knowledge of the things which are lovely and of good report; and that the laws, the life, and the joy of beauty in the material world of God, are as eternal and sacred parts of His creation as, in the world of spirits, virtue; and in the world of angels, praise.'[68]

this tour, beginning at Sallanches on September 9th, 1882. An unused passage of manuscript, dated 'Chamouni, 18th September', is printed in the Library Edition: 'the nearly full moon keeping me awake, not unwillingly, with unclouded light, as she rose above the Dôme, and set over the Breven, while the higher two stars of Orion's belt seemed to pause above the peak of the Aiguille Blaitière, a film of white cloud filling the valley to the south, stretching upwards to Mont Blanc—the aiguilles all silver-grey in moonlight.' *Works*, vol. XXXV, p. 532.

66. *Diaries*, vol. III, p. 1149.

67. Ruskin misquotes himself subtly: in 1870 he had written that 'life without industry is guilt, and industry without art is brutality'.

68. *Works*, vol. VII, *Modern Painters V*, p. 464.

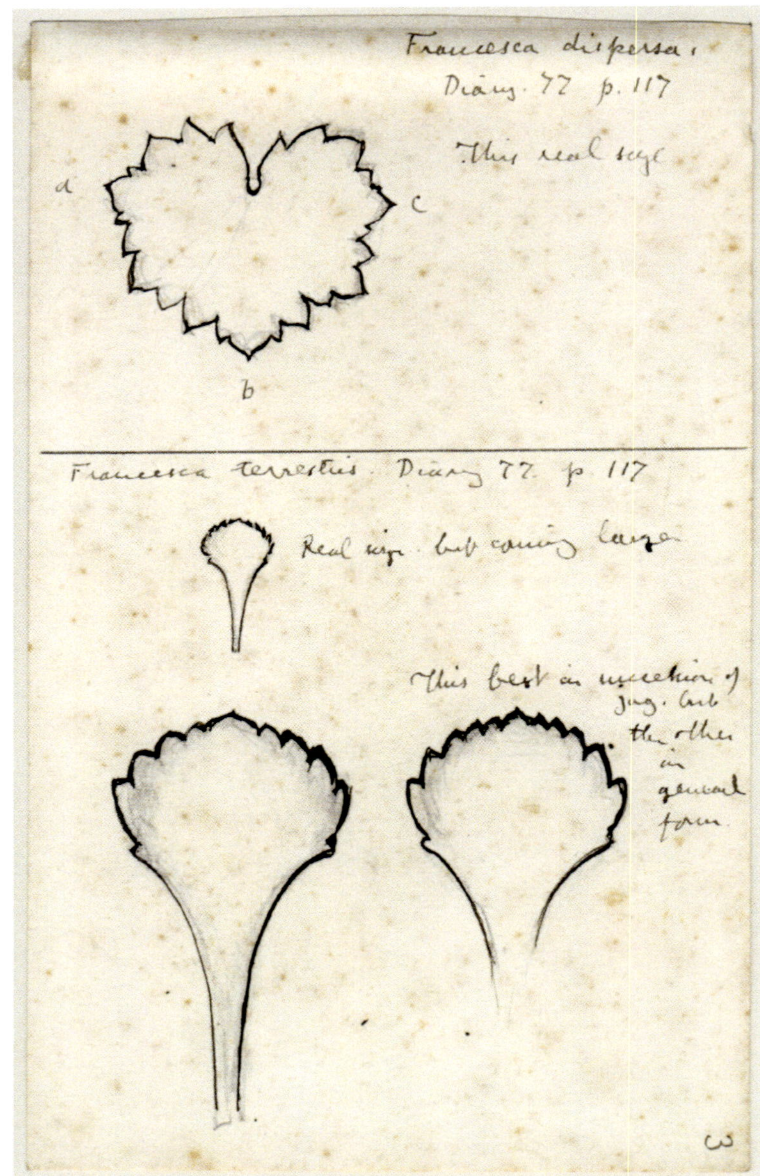

Saxifrages at Brantwood
Left: plants of Saxifraga cuneifolia
Right: John Ruskin, 'Francesca dispersa' and 'Francesca terrestris' [names invented by Ruskin], 1877, pencil and brown ink

ALPINE BOTANY AND THE *FLORA OF CHAMOUNI 1844*

Alpine echoes at Brantwood [1]

Saxifrages

The first Alpine echo to strike me on an early visit to Brantwood, John Ruskin's home in the English Lake District from 1872 until 1900, was the great number of plants of the largely Alpine genus *Saxifraga* between and beside the stone steps leading from the top of the entrance drive to the gardens higher up the hillside. Among the various species and hybrids, including the old garden favourite London Pride (*S.* x *urbium*), were plants of *Saxifraga cuneifolia*, the Shield Leaved Saxifrage, a species abundant in the Alps near Chamonix and included by Ruskin in his 1844 *Flora*.

I think it unlikely that plants of this species were growing at Brantwood when Ruskin acquired the property in 1871, and strongly suspect that he—a dedicated, hands-on and imaginative garden maker—had planted them as a potent visual reminder of earlier Alpine travels.

Precipice Path and Rock (1879)

Another Alpine echo may be found higher on the wooded Brantwood hillside: a steep flight of Ruskin's characteristically uneven stone steps snaking upwards towards a large rocky outcrop forming a miniature cliff or precipice. Beneath it mosses and ferns grow in abundance, although Ruskin originally had another planting scheme in mind, for in his Diary for March 23rd and April 22nd 1879 he wrote, respectively:

> Yesterday making the path up under the 'precipice rock'. I must find less grand name.
>
> Joanie [Joan Severn], on Saturday, completed the planting of my lovely little garden of Hyacinth and Anemone under my 'precipice rock'.

Were the precipice rock and the plantings also intended by Ruskin as reminders of Alpine travels, I wonder, or of visits to Scotland? The answer would depend on what Ruskin meant by 'Hyacinth' and 'Anemone' in his Diary. In the case of the former he may have been referring to one of the blue Alpine *Scilla* (Squill) or *Muscari* (Hyacinth) species, or alternatively to the British *Scilla non-scripta*, usually called Wild Hyacinth in Scotland. In the case of 'Anemone', he could have been referring to one of the many Alpine *Anemone* species, or simply the Wood Anemone, *Anemone nemorosa*, a Ruskin favourite, which grows both in the Alps and Scotland. We may never know, but I like to think that whatever the species planted at its foot, the Precipice Rock itself was intended as a reminder of happy times spent in both the Alps and Scotland.

1. Ingram, David [S.], *The Gardens at Brantwood: Evolution of John Ruskin's Lakeland paradise*, (Pallas Athene, London, 2014); Ikin, Caroline, *Reading Ruskin in the garden: the designed landscape at Brantwood (1871–1900)*, a thesis submitted in partial fulfilment of the requirements of Manchester Metropolitan University for the Degree of Doctor of Philosophy, 2019; Ingram, David [S.] & Stephen Wildman, *Ruskin's Flora: The Botanical Drawings of John Ruskin*, (Ruskin Library and Research Centre, Lancaster University, 2011), p. 29.

Stepped path leading to the Precipice Rock at Brantwood

The Daffodil Meadow at Brantwood

The Alpine Bank, with a fine specimen of a Primula auricula *cultivar next to the rock in the foreground, and a low-growing* Euphorbia *species behind*

Daffodil Meadow (date not known)

A very different Alpine echo may be found near the shore of Coniston Water. There, beside the Harbour Walk (a garden created by Joan Severn in 1899), is a small, uneven meadow which each year for a short season in spring turns gold with early flowering wild daffodils.[2]

Originally these were Lenten lilies (*Narcissus pseudonarcissus*), planted either by Ruskin himself, or by Joan Severn on the instruction of her ageing cousin, to remind him of earlier visits to Vevay on Lake Geneva, where this delightful flower bloomed in profusion.[3] Over the years, however, the Lenten lilies at Brantwood have multiplied and hybridized with local narcissus cultivars and now create a springtime carpet no less golden than their botanically pure predecessors.

2. Ingram (2014), p. 83.
3. Sally Beamish, personal communication.

Alpine echoes planted by Sally Beamish

Other reminders of the Alps at Brantwood were planted more recently by the late Sally Beamish, formerly Head Gardener.

The first is the Mountain Garden (2008), created to celebrate Ruskin's lifelong love of geology and mountain plants. It is situated at the side of the house that faces the lake, and has three principal elements, one being the Alpine Bank, with plantings of *Saxifraga* species and other horticultural alpines such as cyclamens, primulas, gentians and rock rose.

Then, at the top of the entrance drive, close to the house, are six small, faux stone troughs planted to represent different mountain habitats: acid woodland, well drained woodland, tufa rock crevice, acid rock crevice, acid scree and limestone scree. These are valuable in demonstrating the great diversity of soils and plants of the Alps on a miniature scale, but are a poor substitute for the heady experience of actually being in the mountains,[4] as described so vividly by Ruskin in his 1844 Diary.

4. See, for example, the excellent book for lovers of Alpine gardening and botany by Jim Jermyn: *Alpine Plants in Europe—a gardener's guide*, (Timber Press, Portland OR, 2005).

The botany of the Alps

The more one studies alpine plants in a garden, or when walking in the high mountains, the more one craves a deeper understanding of their botany. A good place to begin is the excellent, specialized and comprehensive analysis of Alpine plant biodiversity across Europe found in the volume edited by L. Nagy and others and published in 2003.[5] This publication, together with the volumes listed in the section 'Naming Alpine Plants' (below, p. 35), Jon Mathieu's recent summary of Alpine history,[6] and personal experience, were drawn on in writing the following brief summary.

Zonation and Life-forms

From the earliest times mountains have fascinated and attracted, among others, seekers after religious and supernatural experiences, adventurers, explorers, climbers, artists, writers, philosophers, gardeners, and scientists. Prominent among the latter was Alexander von Humboldt,[7] a polymath whose life straddled the end of the eighteenth and first half of the nineteenth centuries. One of Humboldt's works was, famously, taken by Ruskin as 'light literature', when he went up to Oxford as an undergraduate.[8] Among Humboldt's many and profound contributions to science was the observation that the floras of the world fall into climatic zones that span continents. Thus alpine assemblages of plants, animals and microorganisms may be found across the globe, with the 'alpine zone' occurring anywhere in the world between the altitude where low temperatures limit tree growth, and the permanent snow line, where flowering plants are unable to grow and reproduce. The altitude at which these limits occur depends principally on the latitude of the location. About three per cent of the earth's surface lies in the alpine zone, where snow and glaciers hold about two thirds of the earth's fresh water. It is important to note, however, that the collection of plants made by Ruskin in 1844 in the area surrounding Chamonix, now close to the border between the mountains of the French and Swiss Alps, includes plants from several zones of mountain vegetation, not simply those from the true alpine zone as defined above.

The alpine and mountain plant zones for the present day French Alps, and their altitudinal limits, are approximately as follows:[9]

NIVAL ZONE (ABOVE THE SNOW LINE): in this zone snow lies the year round.

SNOW LINE: from c. 3000 m (north-facing slopes) to c. 3250 m (south-facing slopes).

ALPINE ZONE: to the snow line from c. 2350 m (north-facing slopes), c. 2650 m (south-facing slopes): no trees occur in this zone; the main habitats for plants, which occur in scattered assemblages of prostrate, cushion and rosette forms, include rocks, rock crevices, screes and unstable soils, moraines, and open grassy areas.

UPPER SUB-ALPINE ZONE: to the base of the Alpine zone from c. 2000 m (north-facing slopes), c. 2300 m (south-facing slopes): mainly bushy heathland with ericaceous shrubs (e.g. *Rhododendron ferrugineum* and Bilberry, *Vaccinium* spp.); open grassy areas with other perennial flowering plants.

MIDDLE & LOWER SUB-ALPINE ZONE: to the base of the Upper Sub-Alpine zone, from c. 1450 m (north-facing slopes), c. 1700 m (south-facing slopes): mainly conifer forests (especially: Spruce, *Picea abies* [Lower], mountain *Pinus* species such as *Pinus cembra*, and European Larch, *Larix decidua* var. *decidua* [Middle]); open grassy areas with other perennial flowering plants.

5. Nagy, L., Grabherr, G., Körner, C., Thompson, D. B. A., eds. (2003), *Alpine Biodiversity in Europe*, (Springer, Berlin, 2003).

6. Mathieu, J., *The Alps: An Environmental History*, (Polity Press, Cambridge, 2019).

7. For an excellent summary of Humboldt's life and work, see: Wulf, A., *The Invention of Nature: The Adventures of Alexander von Humboldt—The Lost Hero of Science*, (John Murray, London, 2015).

8. Dearden, J. S., *The Library of John Ruskin*, (The Oxford Bibliographical Society, Oxford, 2012), cat. no. 1365; Wilson, P., '"Over Yonder are the Andes": reading Ruskin reading Humboldt', in *Time & Tide: Ruskin and Science*, ed. by Wheeler, M., (Pilkington Press, Reading, 1996), pp. 65–84.

9. Based on Overal, Bernard, *Guide de la flore des Alpes françaises*, (Naturalia Publications, Turriers, 2016), pp. 9–12.

John Ruskin, Study of a Larch[10] *Bud (a developing female cone),*
pencil, watercolour and bodycolour

MONTANE ZONE: to the base of the Middle and Lower Sub-Alpine zone, from c. 700 m (north-facing slopes), c. 950 m (south-facing slopes): mainly Beech (*Fagus* spp.) or Beech-Silver Fir (*Abies alba*) forests, the mix of these and other tree species depending on the degree of exposure; open grassy areas with other perennial flowering plants.

FOOTHILL ZONE: below c. 700 m (north-facing slopes), c. 950 m (south-facing slopes): mainly deciduous hardwood forests, including Oaks such as Downy or White Oak, *Quercus* spp. and Chestnut, *Castanea sativa*; open grassy areas with other perennial flowering plants; pastures and other cultivated land.

Chamonix lies c. 1034 m above sea level, at approximately the junction of the Foothill and and Montane zones.

The altitudes given above in defining the zones are very approximate and may vary according to location, geomorphology and the aspect of a mountain or valley. Moreover, the margins of the zones are not distinct, with the vegetation characteristic of each merging almost imperceptibly into the vegetation of the zones immediately above and below it. For these reasons there is considerable variation in the altitudes defining the limits of the zones quoted by the several different authors referred to in the 'Collecting Days' chapter of this volume. Also, it is not possible to assign plants in the *Flora of Chamouni 1844* to the zones as currently defined[11] since the upper limits of each zone are probably very different from those of 1844, owing, for example, to normal climatic variation and, more recently, rising temperatures resulting from global environmental change. I have, however, given the approximate current altitudinal ranges for each species in the *Flora*, based on a synthesis of the information provided in the Field Guides listed in 'Naming Alpine plants', below.

10. *Larix decidua* var.*decidua* (syn. *L. europea*).

11. Mathieu, J. (2019), pp. 56–58; Körner, C., Paulsen, J. & Pelaez-Riedl, S., 2. 'A Bioclimatic Characterisation of Europe's Alpine Areas', in Nagy, L., Grabherr, G., Körner, C., Thompson, D. B. A., eds. (2003), pp. 13–28.

At present, in Europe, every c. 100 m increase in altitude is roughly equivalent to 100 km distance north. The steep gradients of high mountains therefore represent compressed latitude-determined zones of plant life, resulting in a relatively rich diversity of species in mountainous regions, especially in the French and Swiss Alps. The European alpine flora—species confined to or predominantly above the tree line—has been estimated to represent c. 20% of European native vascular plants, with very few introduced species as compared with the lowlands. This diversity is extended significantly by variation within species since spontaneous variant populations, maintained or increased by inbreeding or non-sexual clonal multiplication, may persist as a result of their isolation on widely separated peaks. Such intra-specific variation may make the characteristics of particular species very difficult to define with precision and individual plants difficult to name.

Biodiversity has been far less modified by human intervention in alpine areas than in the lowlands, and it is still possible to find some truly wild areas in the alpine zone. Nevertheless, the combined effects of modern tourism, agro-pastoralism, development and global environmental change, including climate change, are placing ever-increasing pressure on the fragile ecosystems in this zone and the zones below it.[12] Interestingly, recent research[13] using large and complex datasets of mountain plant records collected over a period of 145 years from 302 summits across Europe, including the Alps, suggests that the diversity of species on summits has been increasing significantly during recent decades. The rate of increase correlates strikingly with trends in global warming, as distinct from other global changes, presumably as increasing numbers of species are able to grow and reproduce at higher elevations. The ecological significance of such increases has yet to be determined.

The present day French and Swiss Alps, their landscape having been shaped by advancing and retreating glacial action, erosion by river and glacial outfalls, weathering and climate-induced rock falls and screes, exhibit a great diversity of natural habitats for plants. This diversity is increased yet further by human land use, including the creation of high altitude summer grazing areas and changes in the types of animals farmed,[14] and by the mosaic of underlying rocks,[15] from alkaline (mainly limestone or limestone rich metamorphic rocks such as gneisses, with a pH above 7[16]) to acidic (mainly sandstones and intrusive or metamorphic rocks such as granites, schists and gneisses, with a pH below 7). All these factors contribute to the richness of the flora of the Alps.

Winters in the Alps are long, usually with heavy snowfalls from November or December onwards, although the timing, amount and altitudes of accumulations vary according to local, short-term weather patterns and the long-term effects of global environmental change. The winter temperatures of present-day Chamonix and the mountains surrounding the town are likely to be generally higher than they were in 1844,[17] but are quoted in summary below since they give a general sense of the patterns of altitudinal and seasonal change in the region. Thus in the town of Chamonix[18] the current winter temperatures are usually close to zero, but are much lower on the surrounding mountain slopes and peaks: on the summit of the Brévent (c. 2525 m), for example, between approximately -1° C and -10° C in November/December and between approximately -5° C and -10° C in January/February. Even in March/April Brévent temperatures

12. Mathieu, J. (2019), Chapter 8.

13. Steinbauer, M. J. and fifty-two others, 'Accelerated increase in plant species richness on mountain summits is linked to warming.' *Nature*, 556 (2018), pp. 231–234.

14. E.g. Mathieu, J. (2019), pp. 53–56.

15. For a brief summary relating to the area about Chamonix see: Drouet, Serge & Leloupe, Phillipe Hervé, *Guides Géologiques: Vallée de Chamonix—Massif du Mont-Blanc*, (Omniscience, Montreuil, and BRGM Éditions, Orléans, 2015).

16. pH is the negative logarithm of the hydrogen ion activity of a solution or other medium. The pH value (recorded on a scale of 0–14) of a medium gives an indication of its acidity or alkalinity. A neutral medium has a pH of 7; acidic media have pH values lower than 7; and alkaline media have pH values greater than 7. It is important to note that the pH of soils often reflect the pH of the underlying rocks, but this is not always the case, as when a soil comprised mainly of an accumulation of vegetative matter lying over an alkaline rock may have a neutral or acidic pH.

17. Mathieu, J. (2019), pp. 53–56.

18. https://chamonix all year.com/chamonix-weather-guide.

reach no higher than -1° C and may fall as low as -8° C. Also, the effects of low temperatures may be exacerbated by wind chill. Deep snow usually occurs from early November at higher levels and reaches down to or below the level of Chamonix town centre by late December.

The low temperatures inhibit water uptake by plants, resulting in physiological drought, and the slowing down of metabolism. Snow cover excludes most of the light, inhibiting effective photosynthesis, but may also have an insulating effect, providing some degree of protection from both air temperature extremes and the damaging and desiccating effects of wind.

By April the temperatures are rising significantly, especially on the lower slopes, the snow line is receding upwards, and flower buds are beginning to appear on the valley floor. During June the days are much warmer (+18° C in Chamonix but still only +8° C on the Brévent) and the nights remain cool (+12° C in Chamonix and only +2° C on the Brévent). By then the snow has receded to the highest slopes, where it often persists up to the base of the permanent snow line into July.

July and August may be relatively hot in Chamonix itself, up to 21° C, but cooler in the mountains: on the Brévent, +11° C in the day and only +5° C at night. The July and August weather is dry, with the exception of periodic thunderstorms. The thin soils on the mountains exacerbate the effects of dry periods for plants and the relative lack of atmospheric pollution results in a high radiation load on the leaves.

In September conditions remain warm in the town, but plants are already moving rapidly into winter dormancy, particularly at higher levels. By October the nights are cooler, +8° C in Chamonix, but down to -1° C on the Brévent. The leaves of deciduous trees in the lower zones colour up and begin to fall, and the growing season is at an end for almost all plants, even on the very lowest slopes.

Plants of the alpine and mountain zones have evolved various adaptations to help them survive the low temperatures of winter, with persistent snow excluding light, and a short growing season. Physiological adaptations include mechanisms to help prevent disruptive ice crystal formation in the cells and to ensure that the plants are able to endure the desiccation caused by the inhibition of water uptake resulting from freezing temperatures. The life cycle and morphological adaptations of alpine and sub-alpine plants are particularly important in the context of the present volume and partly explain some of the types and groups of plants collected by Ruskin.

As is evident from the specimens included in the *Flora*, most plant species in the Alps, except those that grow at the lowest levels, are perennials. These survive the winter months either as tough, whole plants above ground, or as underground budded rootstocks, fleshy stems (rhizomes) or bulbs. By these means they are able to begin growth as soon as the snow melts, then flower and reproduce rapidly before the temperatures begin to fall again. All the perennial species increase from their vegetative organs, but many also produce fertile seeds, for even on the highest slopes pollinating insects may be abundant,[19] attracted by the unusually bright colours of the mountain and alpine flowers, as Ruskin found in June 1844.[20] Moreover, the parabolic form of some alpine zone flowers (e.g. *Beckwithia glacialis*; *Flora*, Ruskin's pages 34 and 45), concentrates heat from sunlight in the middle of the flowers, increasing their attraction for insects and speeding up the growth and maturation of the sexual structures. Seeds may overwinter to produce seedlings the next season, although these probably do not flower until well established in subsequent years. Other plants may reproduce by forming plantlets or bulbils by non-sexual means in place of flowers but, as with seedlings, such propagules may take some time to form fully established flowering plants.

The tough alpine zone species exhibit a diversity of morphological adaptations to the extreme conditions. Most have a low growth habit: shrubby species usually have a prostrate or ground-hugging form, and vegetative species often have a mat-forming, compact rosette or cushion-like form, with tough persistent leaves. Leaves, moreover, may be hairy or covered with waxy deposits to protect against temperature extremes, high radiation levels and water loss.

19. Like plants, the insect species have their own, distinct physiological and morphological mechanisms to help them survive the extreme winter conditions.

20. See Diary entries for June 13th and 14th, 'Collecting Day Seven', this volume.

John Ruskin, Mountain Rock and Alpine Rose, 1844 or 1849, ink, watercolour and bodycolour

Naming Alpine plants

It may be estimated that there are in the order of 352, 000 flowering plant species on Earth, or c. 391, 000 if other green plants such as conifers, ferns, horsetails, mosses and liverworts are included. Most schemes now used for the naming and classification of this immense diversity[21] are based on the pioneering work of the Swedish biologist Carl von Linné (Carolus Linnæus, 1707-1778), who established the practice of binomial nomenclature, by which all plants are given two names, usually printed in italics. The first is the genus name and is common to all plants in a closely related group. The second is the species name and is unique to a particular type of plant. Further, higher levels of classification show a hierarchy of relationships: genera are grouped into families, families into orders, orders into classes, classes into divisions and divisions into the kingdom Plantae. The prefix 'sub' may be added to existing group names to create intermediate levels of classification, as in the case of subspecies in my Botanical notes in the *Flora*.

The names used in most plant classification schemes are usually derived from the classical languages, normally Latin, but often Greek or Latinised Greek.[22]

The majority of plant classification schemes in the past have been based on morphological and anatomical characters, particularly those associated with reproduction: petals and sepals; stigmas, styles, ovaries, ovules, fruits and seeds; and anthers and pollen. During the late twentieth and the present centuries, however, advances in methods of research, especially numerical approaches and the techniques of genetics and molecular genetics, have made possible the use of other characters. This has led, as will be evident from my Botanical notes in the *Flora*, to some plants being given new species or genus names, or occasionally re-assigned to different families.

It should be emphasized that plant classification schemes are nothing more than cataloguing systems, devised by botanists to help them in their work. A classification scheme does not, therefore, provide proof of relationships between different groups of plants, nor does it necessarily indicate pathways of evolution, but good classification schemes do provide a framework for the discussion and investigation of such concepts.

Plant collectors normally feel a need, indeed an obligation in the case of professional botanists, to identify and name the plants in their collection using existing schemes devised by others, and Ruskin was probably no exception in this regard. This arises, in part, from a desire to extend their knowledge and to find some names as landmarks in the largely unknown field that has interested them.[23] Classifying a collection makes it more manageable as a resource for reference and also requires the collector to examine the specimens closely, thus acquiring an intimate knowledge of their structure and characteristic features. The names and knowledge acquired in this process provide reference points for identifying other specimens that might be collected in the future, either by the owner or another person. Reaching the end of the sequence of collecting, pressing, drying, mounting and naming, moreover, gives the collector a sense of satisfaction, of completeness. Finally, once the plants have been named there is a common language for communicating with others about individual species and plant groups in a collection, or for proposing modifications to existing schemes of classification.

We know that by 1855 Ruskin had rejected all of the then current schemes of nomenclature and classification devised by botanists in the past and was already devising new schemes of his own (See my final chapter in this volume, 'Ruskin and Botany'). However, there is no evidence that he had started to do so in 1844, and indeed he uses conventional Linnæan scientific names throughout his *Flora of Chamouni* for labelling the specimens. Although he identified and labelled many, but not all, of the plants in the collection, he does not reveal what books or other sources he used in making these identifications. The only reference to a possible source in the *Flora* occurs on page 12, where he writes

21. Friday, A. & Ingram, D. S., *The Cambridge Encyclopedia of Life Sciences*, (Cambridge University Press, Cambridge, 1985), pp. 397–416.
22. See: Stearn, W. T., *Botanical Latin: History, Grammar, Syntax, Terminology and Vocabulary*, (David & Charles, Newton Abbot & London, 1990); Stearn, W. T., *Stearn's Dictionary of Plant Names for Gardeners*, (Cassell, London, 2004).
23. Robertson, N. F., 'The Botanic Garden and its Function in Modern Society', *Nature* 194, no. 4823, 1962, pp. 11–13.

'Moerhingia [sic]. G. 223'. The 'G. 223' is possibly a reference to a specific illustration or page number in a botanical book or guide describing a member of this genus, although I have so far been unable to identify what this might be.

Nor does he say whether he made the identifications in Chamonix during the 1844 visit, or subsequently, after returning home. He owned a large library of illustrated botanical volumes by a diversity of distinguished authors,[24] including volumes by Linnæus, Rousseau and Saussure, and Oeder's *Flora Danica*, although many of his books were probably acquired after 1844. He may have inserted the plant names in the *Flora* using such sources, perhaps even long after the plants were collected (but from the appearance of the entries this seems unlikely).

Another possibility is that he consulted local experts during the 1844 Chamonix visit and wrote the identifications on the envelopes he used for storing the specimens when dried (see p. 45, and *Flora* p. 44). Candidates include Venance Payot,[25] who although only eighteen at the end of June 1844 was already acquiring a reputation for botanical expertise, his own very experienced guide Joseph Couttet, or other local Alpine guides.

Finally if, as Stephen Wildman has suggested in his chapter 'Ruskin and Chamonix', Ruskin acquired a ready made collection of pressed and named plants—an *Herbier*[26]—from Venance Payot in 1845, he may have used this, when he took it home later that year, as a reference source for naming plants in his own collection, since by then the specimens would have been mounted in his album entitled *Flora of Chamouni 1844*, with the adjacent blank pages ready for the names and collecting notes. Evidence for this and the previous proposal, admittedly only slight, is that where Ruskin included a colloquial name for a specimen, this was the French name (e.g. page 1 of the *Flora*) and he very occasionally included, in addition to the scientific name, the abbreviated name of the authority for the scientific name (e.g. pages 21 and 28 of the *Flora*), just as Payot did in his *Herbiers*.

Challenges encountered in confirming the names ascribed by Ruskin to the plants included in the Flora of Chamouni 1844, *and in identifying those that were not named*

It has been possible to identify all of the specimens included in the *Flora* to the level of the family, almost all to the level of the genus and the majority of these to the level of the species. In cases of doubt, either because of the limitations of my own knowledge, or the reasons listed below, possible alternative identities have been suggested.

The major challenges to accurate identification of specimens in the *Flora* were as follows.

Many alpine species are morphologically variable, especially given: the diversity of habitat types, each leading to different growth forms; the occurrence of interspecific hybridization; and non-sexual reproduction which may lead to the cloning of hybrids and variants on and in isolated peaks and valleys.

Many specimens in the *Flora* are incomplete, often lacking roots or the whole of the lower parts of the plant, occasionally lacking flowers and rarely including fruits or seeds.

Only one or two specimens of each species were included in the *Flora*, making accurate identification difficult if these were unusual morphological variants.

Ruskin often failed to describe the colour of the living flowers of the specimens, presumably assuming that the natural colour would be preserved.

Many of the specimens are damaged or crushed out of shape, especially the flowers, where important structural features are often obscured. In one case the specimens have disappeared completely.

Because the whole album is a rare object of historical importance, it was not possible to dissect any of the specimens to determine the internal structure of the flowers.

It must be recognized that the naming of plant specimens is in part a subjective process, so error cannot be ruled out in the identities ascribed to the specimens given in my Editor's notes.

24. Wildman, S. (2011) in Ingram, David [S.] and Stephen Wildman (2011), pp. 55–56; Dearden (2012).
25. de Decker-Heftler, S., 'Venance Payot, Guide-Botaniste', an Introduction to *Venance Payot, Herbier des Alpes*, (Les Éditions de l'Amateur, Paris, 2006), pp. 4–14.
26. Ibid.

Sources and conventions used in identifying and naming the plants in this edition of the Flora of Chamouni 1844

Online plant lists

Many on-line plant lists may be used to ascertain the most recent, 'accepted' scientific names of plants, to access images of individal species and to acquire information on, for example, their distribution, threat status, molecular phylogeny and uses. I used two in compiling my commentaries and comments on the plants Ruskin collected in and around Chamonix in 1844, as follows.

The Plant List, versions 1.0 and *1.1* (2013), published on the internet: http://www.theplantlist.org (last accessed September 2021).

This was used in my initial research and at that time was described by the compilers thus:

> *The Plant List* is a working list of all known plant species. It aims to be comprehensive of species of Vascular plants (flowering plants [including rushes, sedges and grasses], conifers, ferns and their allies) and of *Bryophytes* (mosses and liverworts).
>
> Collaboration between the Royal Botanic Gardens, Kew and Missouri Botanical Garden enabled the creation of *The Plant List* by combining multiple checklist data sets held by these institutions and other collaborators [Global Compositae Checklist; The International Legume Database & Information Service; The International Organization for Plant Information; The New York Botanical Garden; The International Plant Names Index; The Convention on Biological Diversity; The Royal Botanic Garden Edinburgh; The South African National Biodiversity Institute; Les Conservatoire et Jardins botaniques de la Ville de Genève].
>
> Version 1.1 (September 2013) replaces version 1.0 which remains accessible via the Version 1.1 website.
>
> *The Plant List* is not perfect and represents work in progress. [The] aims remain to produce a 'best effort' list, to demonstrate progress and to stimulate further work.
>
> Further information is available on the website.

Plants of the World On-line (*PoWo*, 2019). This list, which succeeded *The Plant List*, is facilitated by the Royal Botanic Gardens, Kew, and is also published on the internet: http://www.plantsoftheworldonline.org (last accessed September 2021).

PoWo is easy to accesss and use for both non-specialists and specialists alike, and I therefore made the pragmatic decision to use it in all of my recent work on this edition of the *Flora*. The scientific names given to species described in my commentaries on the plants included in 'Collecting Days' are therefore, wherever possible, in accordance with it. The authority of *PoWo*[27] is testimony to the fact that compiling it and producing many of the data sets on which it draws involves an immense amount of time, effort and dedication on the part of Kew staff and their collaborators.

More specialized plant lists than *PoWo* may be required by those wishing to use the *Flora* as a basis for more advanced scientific and conservation studies, but space constraints preclude detailed consideration of them here.

Books and Floras for identifying present-day French flowering plants

Mabberley, David J., *Mabberley's Plant-Book*, 4th edition, (Cambridge University Press, Cambridge, 2017).

This botanist's *vade mecum* is a treasury of botanical knowledge, the book I always turn to first when looking for information about a plant family, genus or species.

27. It should be noted, however, that scientific names given by me to species in the *Flora* cannot be assumed to have the formal endorsement of the Royal Botanic Gardens, Kew.

Tutin, T. G., Heywood, V. H. and others, *Flora Europaea*, 5 volumes, (Cambridge University Press, Cambridge, 1964–1980, 2nd edition, 1993–).

This comprehensive work, which explores the synthesis of all the national and regional Floras of Europe, aims to be the standard reference, in accessible English, to the names, down to the sub-species level, of all ferns, conifers and flowering plants growing wild or widely cultivated in Europe. It includes keys and descriptions, together with information on the geographical distribution and, where possible, habitat preference and chromosome number. A paperback edition was published in 2010.

Tilson, Jean-Marc & Foucault, Bruno de, *Flora Gallica: Flore de France*, (Biotope Editions, Mèze (Hérault) and the Société Botanique de France, Paris, 2014).

This up-to-date, excellent, French-language Flora covers all wild and widely cultivated species of plants that grow in present-day continental France and Corsica. Produced by a committee of sixty recognized specialists, chaired by Tilson and Foucault, it includes keys to and descriptions of ferns and their allies, conifers and their allies, and flowering plants. Information on phenology, distribution and habitat are also included, together with illustrations of species.

Herbaria

Collections of pressed or otherwise preserved named plants of the Alps are held by national and local natural history museums, universities and botanic gardens throughout Europe. These are of inestimable value for studies of alpine plants. In identifying the plants of the *Flora of Chamouni*, I used the Herbarium of the Royal Botanic Garden Edinburgh, with which I am especially familiar.

Alpine field guides

The books discussed are all intended for use in the field as well as at home or in a library or laboratory.

I used four Alpine field guides while working on the *Flora of Chamouni* and these are listed below, in date order.

I found them all valuable, in different ways, in providing additional scientific and/or visual information about pressed plants included by Ruskin in his *Flora*, as well as for identifying alpine plants, by eye, while walking in the mountains. Some of the scientific names used may be out of date, depending on the date of publication of the guide in question, but names can easily be updated later by referring to a current on-line source such as *PoWo 2019* (see above), which includes historical and current synonyms.

Grey-Wilson, Christopher and Blamey, Marjorie, *Alpine Flowers of Britain and Europe: Collins Pocket Guide*, (Harper Collins, London, 1st edition, 1979; 2nd edition, 2001).

Both versions of this English language field guide are currently out of print, but it is still relatively easy to obtain second-hand copies. Marjorie Blamey's outstanding coloured plant drawings, made from living specimens of almost every plant mentioned, together with Grey-Wilson's detailed and pragmatic descriptions, make these books invaluable tools for visual identification of pan-European mountain flowers and trees, although unfortunately not ferns, horsetails, rushes, sedges or grasses. There is also a short but useful introduction to alpine botany and a concise glossary. The second edition is set out in a more user-friendly way than the first.

Blamey, Marjorie and Grey-Wilson, Christopher, *Cassell's Wild Flowers of Britain & Northern Europe*, (Weidenfeld & Nicolson, London, 2003).

Since a few of the plants included in the *Flora*, such as some Ruskin collected during Sunday or evening strolls, are not true alpine, sub-alpine or montane species, I occasionally used a geographically more wide-ranging book by the same authors as the previous entry. This beautiful book is rather large, so is not ideal for field use.

Ménard, Thierry, *Grand guide de la flore des Alpes*, (Éditions Sud Ouest, Bordeaux, 2007).

This French-language volume, although a little heavy for carrying about in the mountains, fully justifies the effort, for it is a superb field guide covering more than a thousand species of French, Swiss and Italian alpine plants, including ferns, horsetails and their relatives, coniferous and broadleaved trees and shrubs, flowers, rushes, sedges and grasses, grouped according to family. There is a short but valuable introduction to alpine botany, together with a comprehensive and informative scientific description and in most cases a good colour photograph of each species covered. While photographic images are not always as helpful as aids to identification as high quality colour botanical drawings, they do give a very good and accurate visual sense of the habitat occupied by each species, the colour and overall form of its flowers, its key features and its size relative to the plants surrounding it. I found this book immensely valuable.

Hoppe, Ansgar, *A Field Guide to the Flowers of the Alps*, (Pelagic Publishing, Exeter, 2013).

This relatively recent field guide is truly pocket-sized and easily portable, but this is achieved by including only c. 500+ species, about half the number included in Ménard's book. Frustratingly, as the title implies, ferns, horsetails, trees, grasses, sedges and rushes are among the plants omitted. The introduction is also shorter and the technical notes more concise than those of Ménard, although sufficiently informative, when used together with the mainly good colour photographs, for most field needs. A most useful additional feature for non-botanists is that, unlike most botanical field guides, plants are initially grouped according to flower colour, and only according to family within the five broad colour groups (red/pink/orange; white/cream; violet/blue; yellow; and greenish). This decreases significantly the number of pages to be thumbed through by a non-specialist wishing to track down by overall appearance the name of an unknown plant.

Overal, Bernard, *Guide de la flore des Alpes françaises*, (Naturalia Publications, Turriers, 2nd edition, 2016).

This new edition of Overal's French-language volume has a clear focus on the flora of the current French Alps. Like Ménard's guide, it is relatively heavy, but what it loses in portability it more than makes up for in the immense value of its content. In short, it is a superb field guide. It includes 31 pages of useful introductory material, including sections on alpine vegetation zonation and the characteristics of the major alpine plant families. The main body of the book is then taken up with the entries for over a thousand individual species, arranged in family groups, including not only the families of herbaceous plants with petaloid flowers, but also the lycopods, horsetails, ferns, conifers, broad-leaved trees and shrubs, pondweeds, rushes, sedges and grasses. Each species entry includes: the scientific and vernacular name; the family name; a photograph or photographs (the quality varies, but most are good); a concise description; a visual guide to the altitude range of the species and its degree of rarity in each of the six French Alpine departments; and the preferred habitat type. An appendix provides information on alpine vegetation types and groupings. At the end of the volume there is a diagnostic key to alpine plant families, a phyto-sociological classification, a table of species listing, for each one, its degree of rarity in the six French Alpine departments and its phytosociological group, a glossary, and indexes of scientific and French colloquial names.

Alpine geology

Drouet, Serge and Leloupe, Phillipe Hervé, *Guides Géologiques: Vallée de Chamonix—Massif du Mont-Blanc*, (Omniscience, Montreuil, and BRGM Éditions, Orléans, 2015).

This French language guide is a most useful additional resource for any amateur Alpine botanist visiting Chamonix. It begins with a general introduction to the geology of the Mont-Blanc and Aiguilles Rouges massifs. There then follow nine separate suggested itineraries for exploring the geology of the area. Each is well illustrated with photographs and diagrams and mixes geological information with other relevant material, such as the nature of the landscape and the diversity of the flora and fauna which may be encountered along the way. The guide also includes a geological time chart at the beginning, and ends with a short but useful geological glossary and a short but comprehensive and valuable bibliography, including websites.

MAUBERT del.
F. LEBLANC

Collecting, Pressing and Drying Plant Specimens in the 19th century

The equipment of an 1844 collector of plants for pressing was relatively simple,[28] although the individual items may have been bulky or awkward to carry, especially in the Alps. This is unlikely to have concerned John Ruskin, however, for as is clear from his Diary entries for June 1844,[29] and his partial autobiography, *Præterita*, not only was he usually accompanied by a most experienced Alpine guide, Joseph Couttet, but also by his 'man', (John) 'George' Hobbs. The former would probably have been far too grand to be used as a porter, but no doubt George would have been deputed to carry whatever equipment was required.

Herborizing Box or Vasculum

The most important item of equipment would have been a container—a herborizing box or vasculum[30]—for carrying the specimens without crushing them. The term 'vasculum', still in use today, was first coined by Linnæus, in his book *Philosophia Botanica* (Stockholm, 1751, page 293). Among the items he recommended his students to take with them when collecting plants, was a 'Vasculum Dillenianum', a phrase derived from the Latin for a small vessel and the Latinized name of the German botanist, Johan Jacob Dillenius, friend of Linnæus and Professor of Botany at Oxford from 1734 until his death in 1747. The earliest plant collecting boxes of the seventeenth century may have been much simpler, merely being wooden candle boxes, or lighter and more convenient cylindrical, tin-plated metallic candle boxes. The latter were usually fitted with a hinged lid, secured by a hasp and loop, and had hanging lugs, which would have provided anchorage for the two ends of a shoulder rope or strap.

Later, the vasculum became a standard piece of botanical equipment (Allen, 1959) and in Britain was usually an elongated metal box, elliptical in section, tinned or lacquered (usually black or green) inside and out to prevent rusting, its hinged side opening secured with a sliding bolt or some other fixing device (ill. overleaf). The vasculum was often also fitted with a shoulder strap and/or wire handle. Graves (1818), however, suggested an improved 'herborizing box' which had, in addition to a side opening, 'one end … to draw off, by which larger specimens can be admitted without risk of breaking'. European vascula were a little different from those used in Britain, often being cylindrical in section and fitted with two or more compartments (ill. p. 43). Both in Britain and Europe, the more sophisticated versions were sometimes decorated with painted designs.

The early British vascula were very small. For example, the

28. Graves, G., *The Naturalist's Pocket Book or Tourist's Companion*, (Longman, Hurst, Rees, Orme and Brown, London, 1818), pp. 278–301. See also Graves, G., *The Naturalist's Companion*, (Longman, Hurst, Rees, Orme and Brown, London, 1824); Verlot, M. B., *Le Guide du Botanist Herborisant*, (J. B. Baillière & Fils, Paris, 1865), pp. 26–104; Larsen, A., 'Equipment in the Field', in N. Jardine, J. A. Secord & E. C. Spary, eds., *Cultures of Natural History*, (Cambridge University Press, Cambridge, 1996), pp. 358–377.
29. Evans, J. & Whitehouse, J. H., *The Diaries of John Ruskin, Vol. 1, 1835–1847*, (Clarendon Press, Oxford, 1956), pp. 275–296; *Works*, vol. XXXV, *Præterita II*, p. 343 (443) and pp. 328–329 (440–442).
30. Allen, D. E., 'Origin of the Vasculum', *Proceedings of the Botanical Society of the British Isles* (*Proc. BSBI*) 1957, 2, p. 324; Baker, H. G., 'Origin of the Vasculum', *Proc. BSBI*, 1958, 3, pp. 41–43; Allen, D. E., 'The History of the Vasculum', *Proc. BSBI*, 1959, 3, pp. 135–150; Allen, D. E., 'Some Further Light on the History of the Vasculum', *Proc. BSBI*, 1965, 6, pp. 105–109.

Opposite: The fully equipped botanist: frontispiece illustration to Bernard Verlot, Le Guide du Botaniste Herborisant, *1865. Note the gaiters, wideawake hat, long-handled trowel, magnifying glass, and pack for a vasculum and other necessities*

Three British vascula from the collection of the Royal Botanic Garden, Edinburgh. The two black-painted tin ones with side lids only are probably late nineteenth century, the rear one by A. H. Baird of Edinburgh and that in the middle by Baird & Tatlock, at that time of Glasgow, Edinburgh, Manchester and Liverpool. The green vasculum, with a side and end lid, is probably twentieth century, maker not known

Birmingham botanist William Withering, an early proponent, writes: 'Carry them [plant specimens] home in a tin box nine inches long, four inches and a half wide; and one inch and a half deep. Get the box made of the thinnest tinned iron that can be procured; and let the lid open upon hinges.'[31] Such boxes could, therefore, easily be carried in a large pocket. By 1840, however, vascula had become quite bulky, being up to twenty inches long, eight to nine inches wide and five inches deep.[32] Specimens may have been kept fresh while collecting by putting wet moss in the vasculum or lining it with a damp cloth or paper. Indeed, Withering states that; 'If any thing happens to prevent the immediate use of the specimens you have collected, they will keep fresh two or three days in the box, much better than putting them in water.'

It is not known whether Ruskin actually used a vasculum during the 1844 Chamonix visit, but his collecting notes on page 17 (June 10th) of the *Flora* suggest that he used a container of some sort, for he states there that: 'The meadows ... were always covered with the bell gentian and globe Ranunculus ... but I only gathered these grasses [*sic*], for fear of crushing my primulas and

31. Withering, W., *An Arrangement of British Plants: according to the Latest improvements of the Linnæan System, to which is prefixed, an Easy Introduction to the Study of Botany; Illustrated with Copper Plates*, (printed for the author by M. Swinney, Birmingham, 1776), 'Introduction', pp. xlvii–xlxi; Ruskin possessed a third edition of the book (1796), see Dearden (2012), cat. no. 2900. See also Withering, W., *A Botanical Arrangement of all the Vegetables naturally Growing in Great Britain* (1st Edition, printed by M. Swinney, for T. Cadel and P. Elmsley in the Strand, and G. Robinson, in Pater-Noster-Row, London, 1776).
32. Greville, R. K., 'Directions for Collecting Botanical Specimens and Preserving Them for the Herbarium', *Annual Report of the Botanical Society of Edinburgh*, 1838–39, pp. 80–89. Although Ruskin possessed two of Greville's publications—see Dearden (2012), cat. nos. 1130 and 1131—he is not known to have possessed this paper.
33. In my experience everything from sandwich boxes to high crowned hats have at some time been pressed into service to serve as vascula.

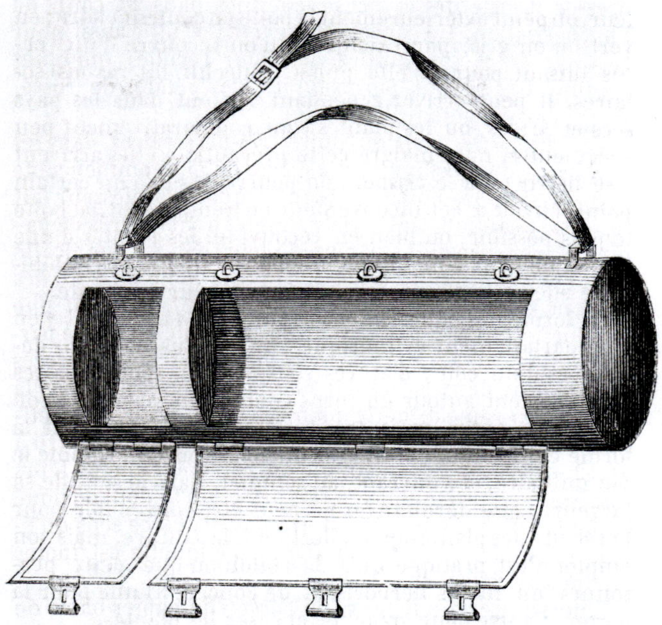

Fig. 9. — Boîte d'herborisation.

A European vasculum—a Boîte d'herborisation—with two compartments, from Bernard Verlot, Le Guide du Botaniste Herborisant, *1865, (Fig. 9)*

saxifrages'. Since the vasculum had become a standard piece of equipment by 1844 it is most probable that he did use one, or if not a container[33] that did duty as one.

The vasculum went on to have a long life: I remember using one while studying botany at school in the late 1950s, and then while an undergraduate in the early 1960s, but by the mid 1960s vascula were being replaced with the more prosaic, but infinitely more convenient, plastic wallet or polythene bag. With concern for pollution of the oceans by plastics, however, perhaps metal vascula will soon make a comeback.

Notebook and Writing Instrument

The second most important pieces of field equipment were—indeed still are—a notebook and a writing instrument, in order to record exactly where and when each plant specimen was collected, the nature and geology of the terrain, the colour of the flowers (for this may be lost during drying) and other details not intrinsic to the specimen itself but likely to be of value both to the collector and other botanists wishing to study the species in question. As Graves (1818) emphasizes, it is this kind of information, which must be written down while still fresh in the mind, that transforms what would otherwise be no more than a souvenir of a day in the field into a scientific botanical specimen.

Ruskin does not appear to have followed Graves' example to the letter, however, for his plant collecting notes were sometimes rather sketchy, even when taken together with information recorded in his Diary for the day in question. Using writing equipment in the Alps in 1844 cannot have been easy, so it is not surprising that one gets the impression when reading Ruskin's collecting notes that they were usually written up from memory, either soon after returning to his lodgings or very much later with the aid of his Diary notes. Sometimes important information was forgotten completely. For example, in the collecting notes on Ruskin's page 11 of the *Flora* he writes: 'I cannot find the name of the plant *c.* and I have lost the ['to' crossed out] name of the locality from which I obtained the other potentilla, *a.*' On page 36, he writes of two specimens: 'Also from the side of La Côte, June 21. I was so fatigued when I came down that evening that I fell asleep in writing; and never noted the exact localities.'

Other Field Equipment

Other equipment in Ruskin's case may only have been a knife or a trowel for digging up specimens, although Verlot (1865) suggests that his botanist should carry a veritable arsenal of equipment, including picks, secateurs, a long handled pruning hook, a walking pole with a trowel at one end, a hand lens or lenses and so on, together, even, with treatments for the various disasters that might befall a botanist in the field, including blisters, strains, stings and bites. It is possible, therefore, that some additional equipment of the kind listed by Verlot may have been carried by the Ruskin party in 1844.

Ordering and Pressing the Specimens

After returning to his lodgings, Ruskin clearly set to work ordering his plants, a tedious task to be performed after a long day in the field, as on June 8th, 1844. The day started at 8.30 am, when he rode first in a charabanc to the mill half way to Argentière, then on a mule for three quarters of an hour, followed by walking and scrambling for the whole day, arriving back at his lodgings at 9.00 pm '– able to arrange my flowers and to sit down now ½ past ten, to write.' Or it was a task to occupy a wet afternoon, as in the Diary entry mentioned earlier for June 27th when 'I had nothing for it but ordering my plants ... I believe I shall have some pleasure out of them.'

A component of Ruskin's 'ordering' of his specimens, apart from writing up his notes, will have been placing them in a press to flatten and dry them prior to mounting. Some botanists actually carried portable presses into the field instead of a vasculum. These usually consisted of two oblong, wooden or wire lattice frames, with many layers of absorbent paper sheets between them, the whole package being bound together with straps. Such devices were cumbersome to carry and awkward to use in the field, the straps having to be removed and replaced every time a new specimen was collected and arranged in the press. The fact that Ruskin mentioned on one occasion that he only gathered grasses, for fear of crushing his flowering plants, suggests that he did not have a portable press with him while collecting in the mountains. However, portable presses were, indeed still are, valuable while travelling, and it is quite possible that he had one or more portable presses in his lodgings and while on the road.

A much heavier type of press, even less appropriate for use in the field but essential in a herbarium and possibly carried by Ruskin while travelling, consisted of two thick, rectangular wooden boards. These would have been carefully selected and seasoned to reduce the risk of warping. The board to be used as the base would have had threaded metal rods attached securely at each corner, or in the centres of the two ends, and these would have passed through matching holes in the upper board, with wing nuts and washers being used to press this board down onto the stack of specimens between absorbent paper sheets supported on the base panel. Withering (1776) describes how to construct such a press and Verlot (1865) illustrates a version with only two tightening screws (see opposite).

Before using the press, Withering (1776) first instructs the botanist to: 'get half a dozen quires[34] of large soft spongy paper; such as stationers call Blossom blotting paper, is the best, and a few sheets of strong paste-board'. Next, to gather the plants of interest on a dry day after the sun has dried off the dew, only selecting specimens 'when their generic and specific characters [notably their flowers and fruits] are most conspicuous', and then to carry them home in the aforementioned 'tin box'. He goes on to describe the use of the press, as follows.

> When you have decided to preserve them [the specimens], suffer them to lie upon a table until they become limber, and then they should be laid upon a paste-board, as much as possible in their natural form; but at the same time with a particular view to their Generic and Specific characters. For this purpose it will be advisable to separate one of the flowers and to display the generic character [the sepals, petals, stamens and pistil]. If the specific character depends upon the flower or upon the root, a particular display of that will likewise be necessary. When the plant is thus

34. A quire was 24 (now 25) sheets of paper, or one twentieth of a ream, unfolded or with a single fold.

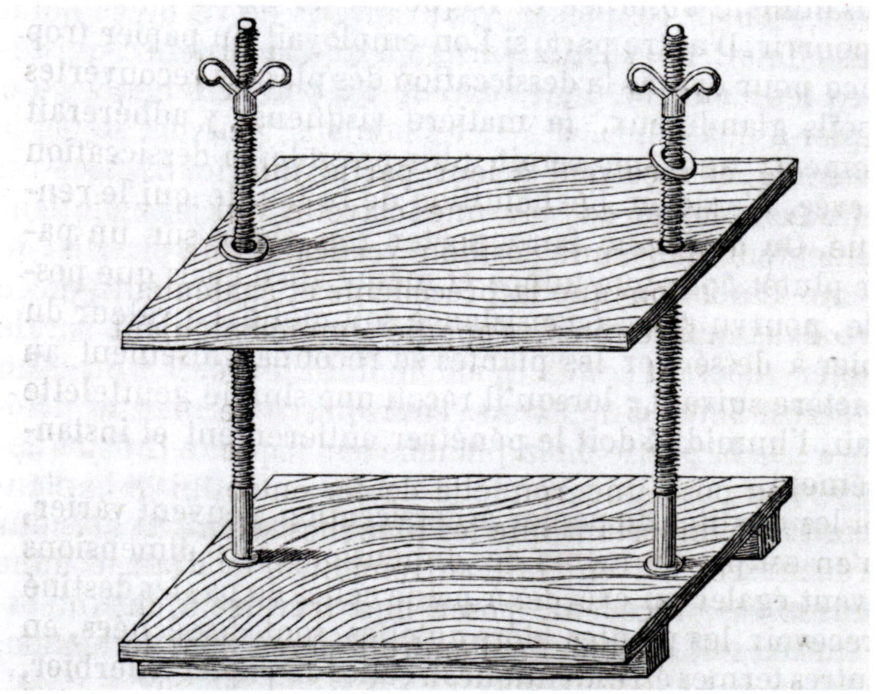

A nineteenth century plant press with two tightening screws, illustrated in Bernard Verlot, Le Guide du Botaniste Herborisant, *1865, (Fig. 10)*

> disposed upon the paste-board, cover it with eight or ten layers of spongy paper, and put it in the press. Exert only a small degree of pressure for the first two or three days; then examine it, unfold any unnatural plaits, rectify any mistakes, and after putting fresh paper over it, screw the press harder. In about three days more, separate the plant from the paste-board, if it is sufficiently firm to allow a change of place; put it upon a fresh pasteboard, covering it with fresh blossom paper, let it remain in the press a few days longer. The press should stand in the sun-shine, or within the influence of a fire.

The use of heat to hasten the drying process is important since rapid drying inhibits the growth of moulds and minimizes discolouration and loss of flower colour. We do not know how easy it was for Ruskin to dry his specimens in this way: sometimes the weather was warm and sunny during the June 1844 visit, but at other times it was wet and cold, so rapid drying might have been a problem. It is possible that, to speed up the drying process, Ruskin used a 'more expeditious' method suggested by Withering:

> Take the plants out of the press, after the first or second day; let them remain upon the pasteboard: cover them with five or six leaves of blossom paper, and iron them with a hot smoothing iron, until they are dry. If the iron is too hot it will change the colours; but some people taught by long practice, will succeed very happily. This is quite the best method to treat the Orchis and other slimy mucilaginous plants.

Somehow I cannot quite imagine Ruskin wielding a hot smoothing iron, but given the frequent wet weather, perhaps Hobbs was pressed into doing so.

In his collecting notes accompanying the specimen of *Globularia cordifolia* L. in the *Flora* (page 4), Ruskin included the words: 'It was I think, much fresher in appearance when I first took it out of its envelopes, and has lost what remained to it of colour by exposure to the light.' This suggests that after pressing and drying his plant specimens, he may have put them into envelopes for safe storage before they were mounted. If he did use 'envelopes' to store some of his dried specimens, as seems possible, it is not known whether, having completed the pressing and drying process, he put the specimens in the envelopes while still travelling, or stored them in a press or presses until he got home and only then put them in envelopes for storage until they could be mounted.

Mounting and Treating the Specimens

Withering later describes the process of mounting the specimens and subsequently treating them to protect against attack by insects, a common problem even today:

> When it [the specimen] is properly dry, the usual method is to flatten it down with paste or gum water, on the right hand inner page of a [folded] large sheet [presumably foolscap] of writing paper. It requires some dexterity to glue the plant neatly down, so that none of the gum or paste may appear to defile the paper. Press it gently again for a day or two, with a half sheet of blossom paper betwixt the folds of the writing paper. When it is quite dry, write upon the left hand inner page of the paper, the name of the plant; the specific character; the place where, and the time when it was found; and any other remarks you think proper. Upon the back of the same page, near the fold of the paper, write the name of the plant, and then place it in your cabinet.

Withering further comments: 'When no better convenience can be had, the specimens may be disposed systematically in a large folio book; but a vegetable cabinet is upon all accounts more eligible.' Ruskin obviously used the less desirable method of mounting his Chamonix specimens in the large album chosen for the *Flora of Chamouni 1844*. It is not known whether this album was purchased in France or England, nor who mounted the specimens, but judging by the quality of the mounting, the job was clearly carried out by an experienced specialist. Whoever this was, it was probably not the person who mounted the specimens in a Payot herbier[35] (see Stephen Wildman's 'Ruskin and Chamonix' in this volume), for the techniques used, although superficially similar, were very different in detail. In both cases, the specimens were not glued down, but having been arranged singly or in groups upon the thick paper leaves, were then fixed down with narrow paper strips, which passed over the specimen. In Payot's album, as on Verlot's herbarium sheets (ill. opposite), these were secured by being glued to the facing surface of the specimen page. In Ruskin's album (ill. p. 48), by contrast, each paper strip was secured by passing the two loose ends through pre-cut slits in the page, and then (probably) glueing them down on the verso of the page. The collecting notes, which included variable amounts of information, were added in ink, in Ruskin's hand, on the blank page preceding the specimen page, sometimes together with the name of the specimen or specimens. Whether these notes and names were added as soon as the mounting was complete, or later, is not known.

Although Withering suggested that specimens should be arranged 'systematically' (by which he meant taxonomically, as by a botanist), Ruskin's specimens are arranged chronologically, according to the date of collection. This enables them to be related to entries in his Diary, which may have been his intention. Unfortunately, this system tells us nothing of Ruskin's attitude to conventional plant classification at an early stage in his development as a botanist. Later, as detailed in my chapter 'Ruskin and Botany' in this volume, he not only rejected conventional plant taxonomies, but devised new schemes of plant classification of his own.

Dried plant specimens mounted on paper herbarium sheets or, as in Ruskin's case, in an album, were, as now, particularly prone to attack and destruction by insects, unless protected in some way. In the eighteenth and nineteenth centuries this was usually achieved by painting ('washing') them with highly toxic chemicals. Withering notes that: 'A small quantity of finely powdered arsenic, or corrosive sublimate [mercuric chloride], is usually mixed with the paste or gum water [used for glueing down the specimens], to prevent the devastations of insects; but the seeds of Staves-acre [*Delphinium staphisagria* L.],[36] finely

35. Payot, V. (2006), *Herbier des Alpes*.
36. All parts of this species contain diterpene alkaloids, which are highly toxic to humans as well as insects. In the past, diluted extracts were sometimes used medicinally.

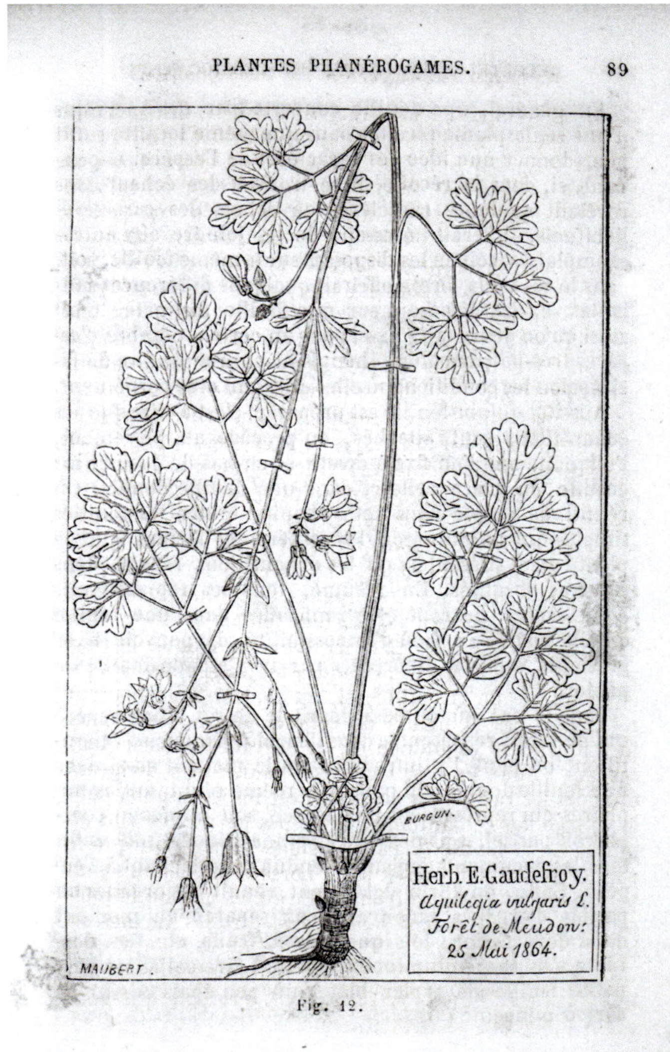

A well mounted and labelled herbarium specimen as illustrated in Bernard Verlot, Le Guide du Botaniste Herborisant, *1865, (Fig. 12). Note that in this example the paper strips securing the specimen are glued to the surface of the sheet*

powdered, will answer the same purpose, without being liable to corrode or to change the colour of the more delicate plants.' Graves writes:[37] 'to avert … attacks [by insects], Sir James Edward Smith (in a letter to Mr. C. Konig, inserted in *Annals of Botany*, vol. II, p. 194) recommends 'a solution of corrosive sublimate of Mercury, in Spirit of Wine [ethyl alcohol]'. With this the learned President of the Linnæan Society [of London], washed his whole collection, and remarks, that since he so did, he has not seen any appearance of insects in the collection; the proportion is about two drachms to a pint, to which may be added a little Camphor.'

The plant specimens in Ruskin's album are remarkably insect free, so it is likely that he did treat them with insecticidal toxins in some way, but what these were can only be discovered by chemical analysis. In the absence of such analysis, it is suggested that the album and specimens are handled with extreme caution.

37. Graves (1818), pp. 297–298.

Ranunculus glacialis *(now* Beckwithia glacialis, *Glacier Crowfoot),*
the final pressed plant included in Ruskin's Flora of Chamouni 1844

The Collection of Pressed Plants made by John Ruskin in the Alps near Chamonix in 1844

Motives for making the collection

Ruskin gives no reasons for deciding to make a collection of pressed plants[38] during his visit to Chamonix in 1844, either in the *Flora* itself, or in his Diary for 1844. Yet make it he did, at considerable cost in terms of the time spent in arranging, pressing and drying the plants, then mounting them (or more probably, paying to have them mounted, no doubt at great financial cost) in a handsome green, calf-bound album, fitted with a brass lock and decorated and lettered in gold on the front cover with the grand title 'FLORA OF CHAMOUNI. 1844.' After this, yet more hours would have been expended in identifying many, though not all of the specimens, labelling in his 'best hand' those that had been identified and inserting his sometimes detailed collecting notes.

It was, of course, the thing to do during travels at that time, not least in Chamonix, where the range of colourful, beautiful and unusual plants flowering on the mountains, even as early as June, was considerable. Indeed, to save tourists the trouble of making their own collections, some local guides, for example, Venance Payot,[39] who turned compiling herbiers into a local cottage industry, were willing to sell them off-the-peg collections with the plants already identified and labelled. As Stephen Wildman has pointed out in his chapter on 'Ruskin and Chamonix', Ruskin may have ordered one of these from Payot, to be collected during a flying visit to Chamonix the following year. It is clear, however, that the *Flora* is not Payot's work. Firstly, because Ruskin, in his collecting notes, often supported by the corresponding Diary entries, describes where and when he collected most of the specimens included in the *Flora*. In those cases where he forgot the place of collection, he admits this. Secondly, Ruskin's specimens in the *Flora* are mounted in a different way from those in Payot's herbier (see above). One therefore returns to the question of why he should have made a collection of his own.

He may have made it as a form of botanical self-instruction, requiring him to observe the form, location, degree of rarity and, above all for an artist, the colour of the plants and flowers, and later to practise naming them, referring to appropriate texts. If this were the case, however, it is strange that the choice of plants collected was so random (see below) and that some of the collecting notes had such sparse information of this kind, apart from comments on the colour of the flowers, including, in the case of the Gentian on Ruskin's page 4 of the *Flora*, the note that 'it has not only lost its colour, but shrunk to one half its original size, and is now a very slander on itself and its companions ...'

It is easier to suggest cogent reasons why the naming of specimens was abandoned before all had been identified: perhaps appropriate sources of information were not available, or the process of identification was too difficult or, most probably for Ruskin, some new and more interesting project took first his attention and then his energies before the *Flora* project was complete.

One can only conclude that for Ruskin the plants collected and pressed in Chamonix in 1844 were not seen as part of a separate, systematic botanical study of the Alpine flora, but as an adjunct to his Diary, visual *aides mémoires* to serve in lieu of detailed flower sketches, drawings and descriptions made on site or from memory after a long, tiring day in the field. They would thus serve as integral parts of Ruskin's comprehensive record of the 1844 Chamonix visit, alongside his detailed descriptions, sketches and drawings of cloud formations, the sky

38. A 'herbier' or 'herbarium'.

39. Payot, V., *Herbier des Alpes* (Éditions de L'Amateur, Paris, 2006), pp. 5-15.

and the terrain, and his accounts of the weather. By 1844 he was perfectly capable of making flower drawings, noting much later in the Introduction to *Proserpina* that 'I had begun my studies of Alpine botany ... in 1842, by making a careful drawing of wood-sorrel at Chamouni ... '. But to make such drawings would have required considerable time and would have slowed the progress of each day's expedition to a snail's pace, or would have greatly extended each evening's labours in writing up his Diary; it would have been easier to put the specimens into a container or plant press to be dealt with later. Credence for this suggestion may be derived from the fact that the specimens, when arranged in the grand album, were not ordered according to accepted botanical principles, or colour, or any other physical feature, but as noted above, by the date of collection, which corresponded exactly with the Diary entries for the day. Indeed, considerable care was taken to record the exact date and place of collection, even with a specimen deliberately mounted out of sequence for dramatic effect, as in the case of a magnificent specimen of *Ranunculus glacialis* (*Beckwithia glacialis*) (see ill. p. 48). This was mounted on Ruskin's page 45, the last page of the *Flora*, after a series of specimens collected on June 27th, despite having been 'gathered June 22nd', because it was 'most thoroughly characteristic of the Alps'.[40]

Other circumstantial evidence supporting this conclusion is that the only plants actually drawn and described in detail in the Diary during the 1844 visit were trees, which could not be pressed. In these cases leafy branches could be taken back to Ruskin's lodgings to be drawn at leisure, without losing valuable expedition time.[41] For example, on June 16th Ruskin sat in his room with 'a branch of sapin' beside him, presumably gathered the previous day 'from one of the topmost fragments of the shattered pines of les Pelerins'. Having drawn the branch, he then wrote detailed notes about it. Then, on June 20th, despite the beauty of the early morning view, it was 'mighty cold', so having failed to draw the view he 'sat down to [his] sapins again', these being the branch or branches he described and drew on the afternoon of Sunday June 16th.

It is not known why Ruskin did not make further collections of Alpine plants flowering at different times of year on subsequent visits to Chamonix. One possibility is that he found the whole process of pressing, drying and mounting the specimens so involved that he did not feel that it would be a good use of his time to make another collection. A second possibility is that he was so disappointed that the flowers in the first collection all lost their colour that he could not see the point of repeating the exercise. A third, more likely possibility is that he was able to obtain a ready made, fully labelled collection which covered the whole flowering season in the area, from Venance Payot, as suggested by Stephen Wildman in 'Ruskin and Chamonix'.

The constraints limiting the size and scope of the Flora of Chamouni 1844 *collection*

The *Flora* cannot be regarded as a comprehensive representation of the flora of the Alps of the Chamonix region, not least because it was limited to the month of June 1844, when many plants would not yet have come into flower after the long, hard Alpine winter. The scope of the collection, therefore, was limited to spring flowering species.

A further constraint was the weather, which was very variable and often too wet for serious collecting, especially in the second half of the month. For example, June 19th was not a good day according to the Diary, the rain having 'plashed ponderously' all the previous night, and although the sun was out when Ruskin rose and at 10 am almost enticed him out to draw, he luckily stayed in to work on his plant collection, for the rain then set in for the day. He made a dash to collect some plants later in the day, but was caught by the rain and no specimens found their way into the *Flora*. He ended the Diary entry by noting that the mountains were by then 'covered with fresh snow'. On June 25th it rained in the morning, but half cleared at 10 am, so Ruskin took a charabanc and 'botanized ... a little', but was 'again driven home by rain.' Whether any plants were collected is not

40. *Flora*, Ruskin's page 45.

41. See Diary entries for June 12th and 16th, 1844, pp. 93–4 and 103 of this volume.

known, but none survived to be included in the *Flora*. June 26th started well at 4.15 am, but the weather again deteriorated rapidly into 'a disagreeable windy bleak day for drawing: worked hard and spoiled all as usual.' In a short ramble to warm himself he found 'a glorious little bit of scenery by the Arve side', and had 'a delicious climb … with Couttet', but once more no plants appear to have been collected. June 27th was also very wet, with Ruskin reduced to 'ordering [his] plants in the afternoon.' Despite the rain, however, three plant species, albeit from the valley rather than the mountains, were collected for the *Flora* at some point during the day.

Some days were, by contrast, extremely hot, although heat did not necessarily lessen Ruskin's urge to collect. For example, although June 10th 'turned out a day of awful heat …', he managed to acquire seven plant species for the *Flora*. June 24th was also 'Desperately hot. … the atmosphere like that of an oven', but despite the heat and other challenges the day yielded three interesting and contrasting plant specimens, which were ultimately mounted together on a single page of the *Flora*.

Further time was lost in caring for his parents. For example, the Diary entry for June 8th records that having rested until 7.00 am, the Ruskin party was on the road by 8.30 am, first by charabanc[42] to the mill half way to Argentière, where John showed his father and mother the avalanche fallen from the Aiguille Verte; only after this was he able to begin his ascent of the mountains on a mule and on foot. On June 17th he set off at about 8.00 am with his father and mother to show them the fall of the Eau de Bérard, but 'they were neither of them well, lost their tempers and spoiled it all.' The rest of the day, fortunately, was spent more profitably, high above the valley d'Entragues.

Sadly there was to be no more plant collecting after June 27th during the 1844 visit to Chamonix. The weather remained unsettled, and the time was taken up with short outings, sometimes to draw, between spells of rain, and with family cares, for Ruskin's father and mother were both unwell at times. On July 3rd, the final day in Chamonix, he was forced to admit, with a heavy heart: 'The weather has been against me this last fortnight.'

42. A charabanc was an uncomfortable horse drawn vehicle, with bench seating arranged in rows facing forward and open to the air.

An additional limit to collecting was that the Ruskin family regarded Sundays as days of rest or religious activity; however, John managed to squeeze a little botany into some of them. For example, on Sunday June 9th, after spending the morning 'reading divinity', he went out for a 'saunter', then after dinner for a 'ramble', and managed to collect three low altitude species for the *Flora*. On Sunday 16th June he at least managed to draw and describe his 'sapin', as already discussed, but on Sunday June 23rd he achieved little more than a 'short saunter … in the forenoon' and a 'meditative walk' in the afternoon, with no plants being collected for the *Flora*.

A final constraint was the wish to paint ('draw') the Alpine scenery and sky as well as to collect plants. Thus on Saturday 15th June it was too hot to climb, so Ruskin took the charabanc, and after walking through a field or two with his father sat drawing near Bossons for three or four hours, impeded only by breezeless heat, and no plant specimens were collected for the *Flora*. On June 18th he spent much of the day drawing, getting on 'pretty well, though interrupted by … light showers'. Once again, no plants were collected. The drawing did not always go as well as he hoped: on June 27th (see above), when the rain began to pour down, instead of drawing at the avalanche there was 'nothing for it but ordering my plants in the afternoon …', a clear indication of where his priorities lay.

Criteria used for selecting which specimens to collect for pressing

Ruskin provides no account of the factors affecting his choice of plants, so it is necessary to fall back on circumstantial evidence. The apparent randomness of the collection suggests that principally he collected those specimens whose flowers caught his eye, especially because of their colour. Nevertheless, he does seem to have had some ambition to be systematic, for he writes in his notes for June 10th of a collection of grass-like plants: 'These grasses [*sic*] … were intended to form part of a series of grasses; but I was not again among the low meadows, and got only hill plants.' He also may have had the intention of including mosses in the collection, for on June 11th he gathered specimens of Common Haircap Moss and Great Golden Maidenhair Moss, but perhaps it was their habit of growing 'in deep & thick masses' in the woods that caught his eye, rather than any regard for their systematic position in the plant kingdom.

Colour definitely seems to have been of particular importance to him when deciding what to collect. It has already been mentioned that he regretted the loss of colour in the Gentian specimen collected on June 5th. His collecting notes for the same day concerning another specimen, the delightfully named Heart-leaved Globe Daisy (*Globularia cordifolia*) tell us even more. 'It was growing in masses which made the ground purple all about the Lac Benit; associated with the Soldanella. Their delicate purples opposed to the gold of the Potentilla, and the common crowfoot. But its colour is now entirely gone… What vestige of hue remains to it, is however its original one, it has not turned purple from red or blue, as is the case with many plants.' On June 12th he encountered the magnificently coloured, daisy-flowered Golden Hawk's-beard (*Crepis aurea*) growing in steep pastures under the Cascade des Pèlerins and writes in his collecting notes: 'It is a common flower, but I never saw it of so vivid an orange as at this spot.' And then, once again with obvious regret: 'Its colour was quite lost in drying.' On June 22nd, as he ascended above the Glacier des Bossons, he encountered a diminutive plant with an ugly name, Alpine Toadflax (*Linaria alpina*), but delightful flowers and writes: 'The soil … was dry & dusty … & the only redeeming point of the whole was, that this flower, then new to me, — was growing in large bunches of the most intense purple — all about the crannies of the loose soil. The second time that I found it was on the very highest turn of the Simplon road … there also growing on a dry, dusty soil to which its roots hardly adhered. There I found two species [forms], one like this, with the vivid orange centre, the other having this orange replaced by a pale lilac.'

He also seems to have valued the ability of particular plants to grow at high altitudes, writing of the Alpine Toadflax mentioned above, for example, with implied wonder, that 'In both these situations the plant was growing at more than 6000 ft above the sea.' Moreover, altitude linked with obvious hardiness was even more highly valued. Writing of Alpine Snowbell (*Soldanella alpina*), which he had collected on June 5th, he says in his collecting notes, which must have been added long after that date, that three days later, on June 8th: 'on the crest of the ridge which separates the valley leading to the Col de Berard, from the Val d'Entragues, and which overhangs the latter in an awkward precipice, facing north, in crossing some slopes of half frozen snow—I saw the purple head of this flower thrust up through the snow, which it had pierced, though an inch thick.' Similarly, in his Diary for June 17th he writes of the same species: 'To-day high up above the val d'Entragues, Coutet pointed out to me a flower of the Soldanella which had actually pierced the snow, and was budding in the midst of it.' And of Alpine Rock-jasmine (*Androsace alpina*) he writes on June 22nd in his collecting notes: 'This plant is certainly one of the hardiest of the Alps. I found it growing in large bunches, June 22nd, among the loose blocks of the uppermost moraine of the glacier des Pèlerins, at a height of nearly 9000 ft above the sea, close to the ice, and associated only with the Ranunculus Glacialis.' Finally, of this latter species he again writes in his collecting notes for June 22nd that it grows: 'in the most desolate places, with nothing but the moraine sands, wet with ice streams, for a soil. It haunts the borders of the glaciers in a zone of about from 5 to 9000 ft above the sea, seldom occurring lower, nor far from ice…'.

Rarity was a further criterion which Ruskin may have used to select plants for inclusion in the *Flora*. Although collecting rare natural history specimens is frowned upon and often illegal today, in the nineteenth century it was seen as normal or even desirable.[43] Thus in his collecting notes for specimens of Mountain Everlasting (*Antennaria dioica* [formerly *Gnaphalium dioicum*]) acquired on June 9th Ruskin writes, with regret: 'I took them for a rarer flower and had a scramble for them, swinging myself up by the pines'. On June 10th he again regrets that a species was not rarer, writing: 'Josef [Couttet] … brought me

43. Jardine, N., Secord, J. A. & Spary, E. C., *Cultures of Natural History* (Cambridge University Press, Cambridge, 1996).

these junci [Field Wood-rush; *Luzula campestris*] as if they were rare. I kept them therefore, though they seemed to me one of the most frequent meadow plants.'

June 22nd, however, finds Ruskin writing with pride, it would seem: 'This [Purple Saxifrage; *Saxifraga oppositifolia*] I believe to be an exceedingly rare plant. I never saw it but once, and then I could not get at it myself, but Coutet scrambled down into the chasm & brought it up for me, indeed he was the first to point it out.' In fact, it is not particularly rare now, although it may have been so in the Chamonix area in 1844. Even if it was not rare, Ruskin was not to know and he did gain much satisfaction from its colour, which was 'when fresh, ... an intense smalt—a mixture of lake and ultramarine; exquisitely beautiful'.

Finally, and perhaps surprisingly, sentiment and respect seem to have been factors in deciding what to include in the *Flora*. For example, on June 5th he wrote only the briefest of collection notes for two specimens of a Broomrape (*Orobanche*) species, as follows: 'The two specimens were gathered by my Mother, from the meadows under the Nant d'Arpenaz, while I was absent on the Mont du Reposoir.' These dramatic, pale yellow, almost translucent plants lack the green pigment chlorophyll and are therefore dependent for their supply of carbohydrates on their ability to parasitize other plants, yet Ruskin makes no comment on any of these characters. He clearly included them in the *Flora* for no other reason than that they had been collected by his mother. More robustly, in his collecting notes for June 10th, he writes of a sedge specimen: 'Below the saxifrage places, we came to some steep meadows, sloping to the brink of the eau noire [*sic*], where I found this primula, and I imagine the grasses below—Coutet was always gathering grasses; as if I had wanted a hay rick instead of a Flora, and these are probably his selection.' In this case, as in the case of the Field Woodrush Couttet brought him on June 10th, respect for his guide seems to have been a perfectly acceptable criterion for keeping the specimens.

John Ruskin, Star Gentian [possibly Spring Gentian, Gentiana verna *L.].*
Image from the Diary notebook of 1849, 'Star Gentian carried to
Chambery and drawn there 3rd May'

COLLECTING DAY ONE: JUNE 5TH, 1844

From the Diary: May 14th–June 5th, 1844[1]

After a long and arduous journey from Dover, including a final hair-raising descent of the Jura with a drunk postilion in charge, the Ruskin entourage, including a somewhat downhearted, twenty-five-year-old John Ruskin, finally reached Geneva on May 31st 1844. At last they were within striking distance of Chamonix, and on June 1st Ruskin's spirits begin to lift and he enthusiastically recalls his first 1844 encounters with the rich and exquisitely beautiful Alpine flora:

'The day before [May 30th] I should remember, for the walk I had at St Laurent ... The Dôle rose in a long red ridge ... above a range of pine[2] covered hills ... About lay the grey concave blocks of the Jura limestone, slippery with wet. Large black and white snails had come out everywhere to enjoy the rain. In the crevices of the rocks the lily of the valley[3] grew profusely, accompanied by the wild strawberry and cowslip.[4] I found a root of the star gentian[5] [ill. opposite], and kissed it as a harbinger of the Alps.'

The next morning (May 31st), he recalls, just beyond Les Rousses, he 'had a fragment of a lovely walk ... among the gentians. I never saw them in such profusion before nor of such blue: it was as if Heaven had been left desolate and grass had grown on it.'

By June 5th the Ruskins were very close to Chamonix, and John Ruskin had started to collect flowers for his Flora, writing in his Diary on that day:

'ST. MARTIN'S.[6] Two of the most delicious and instructive days I ever spent in my life. Note, first, of the two purple flowers dried today, a round bushy one,[7] grows

1. This and all subsequent 1844 Diary summaries and quotations are based on the manuscript Diary for 1844 held in the Whitehouse Collection, The Ruskin, Lancaster University (MS4) and on Evans, J. & Whitehouse, J. H. (1956), *The Diaries of John Ruskin, Vol. 1, 1835–1847* (Clarendon Press, Oxford), pp. 275–296.
2. Presumably a generic term for conifers of various sorts.
3. *Convallaria majalis* L.
4. Respectively, *Fragaria vesca* L. and *Primula veris* L.
5. The species shown opposite, which Ruskin labelled as 'Star gentians' in 1849, appears to be what is now usually known as Spring Gentian (*Gentiana verna* L.), which flowers from April/May until August, and has five large, clear blue petal-like flower lobes.
6. Presumably at Pont de St. Martin, close to Sallanches, c. 27 km from Chamonix.
7. Ruskin's page 1 of the *Flora of Chamouni*: 'GLOBULARIA CORDIFOLIA' (sic).

in enormous quantities near and about the top of Montagne Bénit; the other more scattered,[8] though in profusion, near the top, and more sparingly half way down the oxalis acetosella,[9] in considerable quantities under the rocks, and in the woods, about two thirds of the way up; the anemone nemorosa[10] more sparingly a little lower. The yellow flower,[11] of which only one bad specimen is dried, in vast quantities high and low. On the summit itself the bell gentian[12] grew in mossy knots, everywhere, of luxuriant size, scattered, though not in such profusion as far as half way down. The star gentian very rare at the top, thickly scattered below with the other.[13] Common crowfoot, daisy and dandelion,[14] up to the top. In the pastures below, the globe ranunculus[15]—yellow—mixed in profusion with the purple orchis.[16] The white plant, No. 1,[17] very far down in the woods; No. 2[18] higher up.'

Editor's botanical note

The area Ruskin visited on June 5th would have been composed mainly of limestone, a calcareous rock. The rocky soils would, therefore, probably have had an alkaline pH (above 7), although in largely organic soils made up of accumulated dead plant remains, as in wooded areas or deep, sometimes boggy hollows, the pH may have been neutral (pH 7) or even acidic (below pH 7). These patterns are reflected in the pH preferences of the eight species, one collected by his mother, which represent the expedition of June 5th, 1844 in his *Flora of Chamouni*.

8. Ruskin's page 3: 'SOLDANELLA ALPINA' (sic).
9. No pressed specimen.
10. *Anemonoides nemorosa (*syn. *Anemone nemorosa* L.*)*,, Wood Anemone; no pressed specimen.
11. Ruskin's page 2: 'DRABA AZOIDES' (sic).
12. Verso of Ruskin's page 4: 'GENTIANA ALPINA' (sic).
13. Presumably verso of Ruskin's page 4: 'GENTIANA ALPINA' (sic).
14. Respectively: *Ranunculus* sp. (a Buttercup); *Bellis perennis* L. (Daisy); a member of the complex group now referred to as *Taraxacum sect. Taraxacum* F. H. Wigg; no pressed specimens of any of these.
15. *Trollius europaeus* L. (Globeflower); no pressed specimen.
16. Possibly *Orchis mascula* (L.) L. (Early Purple Orchid); no pressed specimen.
17. Presumably Ruskin's page 1: 'MYANTHEMUM BIFOLIUM' (sic).
18. Probably Ruskin's page 3: 'Gnaphalium dioicum' (sic), in which the male flower-heads of this dioecious species appear white.

Transcript of Ruskin's page 1[19]

MYANTHEMUM BIFOLIUM
Muguet a deux feuilles[20]

The two specimens, a & b,[21] were gathered on the 5th June,[22] among the woods which clothe the lower slopes of the Mon (du Reposoir?)[23] which rises on the right in approaching Cluse from Bonneville by the old road. The plant seems to love sloping banks, growing chiefly beside the hollow paths cut or sawn out of the gravelly soil by the dragging down of the pinetrunks. These, to the number of 11 or 12, are bolted together by the extremities, and are drawn down the steep sandy slopes by a single horse.

Vide large collection, Vol 3. 238.[24]

Editor's note on specimen

'MYANTHEMUM [SIC] BIFOLIUM' *Maianthemum bifolium* (L.) F. W. Schmidt (May Lily)
Asparagaceae (syn. Convallariaceae; Asparagus family)

May Lily was once thought to belong to the large Lily family (Liliaceae), but as a result of analysis using modern technology has recently been re-assigned to the family Asparagaceae, a group which also includes Wild Asparagus, Lily of the Valley and Solomon's Seal.

This delicate plant is a relatively low, creeping (by means of underground stems called rhizomes), hairless perennial, which often carpets the ground on hillsides at altitudes of c. 1000 to 2100 m. It prefers shady, usually wooded, slightly acid soils[25] (pH lower than 7) rich in organic matter. Each stem normally bears two small scale-like leaves at its base and higher up, two alternate and much larger, smooth-edged, heart-shaped leaves. The white flowers, which appear from May–June/July, are borne in a short spike (raceme) at the tip of the stem and are fragrant. They are relatively small (c. 4–6 mm diameter), with four tepals (petal-like structures which function as both sepals and petals) which, when open, resemble a reflexed four-pointed star. The fruits are small red, spherical berries, c. 4–5 mm diameter.

19. In transcribing Ruskin's collecting notes on this page and all subsequent pages, I have left Ruskin's spellings of all words, including place names, as they were written, unless indicated otherwise in square brackets. Also, I have not inserted any new words or punctuation marks, unless indicated otherwise in square brackets. I have, however, omitted some pen marks where it was clear to me that these were definitely not intended as words or punctuation marks. Where I was uncertain whether a mark was intended as a comma or a full stop, I have used my judgement in deciding which would be most appropriate, given the context.

20. = Lily of the valley with two leaves.

21. Not actually labelled a & b.

22. See also Diary for June 5th, above.

23. Possibly Chaîne du Reposoir, but may in fact be Chaîne du Bargy.

24. It is not known which 'collection' this refers to.

25. Although the place where this plant was collected is in the limestone region of the Alps, the largely organic soil of the woodland floor may well have had a slightly acidic pH.

Transcript of Ruskin's page 2

DRABA AZOIDES

Drave faux. aizoon[26]

This plant was growing, June 5th, in enormous quantities about the Lac Benit and on the bare turfy summits under the great precipice of the Mont du Reposoir, associated chiefly with the Soldanella, Potentilla, & large bell gentian. Vide notes in journal.[27] I found it afterwards frequently in the same situations—but never in such profusion

Editor's note on specimens

'DRABA AZOIDES' *Draba aizoides* L. (Yellow Whitlow Grass) Brassicaceae (sometimes referred to as Cruciferae; Cabbage family)

This bright flowered, low, tufted, hairless perennial prefers calcareous soils,[28] in rocky or stony habitats at altitudes of c. 2500 to 3600 m. The leaves, which are clustered in a basal rosette, are linear to lanceolate in form and edged with stiff bristles that point towards the tip. The yellow flowers, which appear from March to May/June, are borne in clusters on upright stems (c. 10 cm). Individual flowers are tiny, c. 5–7 mm diameter, and have four oval petals arranged in the form of a cross (hence the alternative family name). The whole flower head elongates as the fruits (capsules; not included with specimen) develop and dry out, each capsule having the form of a beaked ellipse, c. 5–12 mm diameter.

A specimen of this species appears in Venance Payot's *Herbier des Alpes*,[29] p. 11.

PINGUICULA

Below the specimen of *Draba azoides* are two unlabelled specimens of Butterwort (*Pinguicula* spp.), members of the Lentibulareaceae (Bladderwort and Butterwort family of insectivorous[30] plants). The specimens are not identified or described, however, until Ruskin's page 30 of the *Flora* (q.v.),[31] where it becomes clear that they have white flowers and are therefore *Pinguicula alpina* L. (Alpine Butterwort).

26. = *drave*, a plant from the Cabbage family + *faux aizoon*, a plant resembling but not belonging to a member of the tropical and subtropical Stone-plant family (Aizoaceae).
27. See Diary for June 5th, above.
28. A soil with a high content of calcium carbonate, normally with a pH of 8 or higher, as would have been the case about Lac Bénit.
29. V. Payot, *Herbier des Alpes* (Les Éditions de l'Amateur, Paris, 2006).
30. Plants which derive some of their nutrients by digesting the bodies of trapped insects.
31. See also Diary for June 17th, below.

Transcript of Ruskin's page 3

SOLDANELLA ALPINA

This, certainly the boldest, though not the most ambitious, of Alpine flowers, was growing, June 5th, in profusion about the Lac Benit & more sparingly half way down. Above the Flegere,[32] June 8th I found them finer, though less numerous—one of these is preserved a[t] p 9.[33] Again, on the crest of the ridge which separates the valley leading to the Col de Berard, from the Val d'Entragues, and which overhangs the latter in an awkward precipice, facing north, in crossing some slopes of half frozen snow—I saw the purple head of this flower thrust up through the snow, which it had pierced, though an inch thick.

Transcript of specimen page

a.	b.
c	d.

a. b. Soldanella alpina c. d. Gnaphalium dioicum.

Editor's note on specimens

'SOLDANELLA ALPINA'

'a. b. Soldanella alpina' *Soldanella alpina* L. (Alpine Snowbell)
Primulaceae (Primrose family)

These elegant Alpine perennials, which usually form large groups, prefer calcareous soils[34] in moist pastures and stony areas, frequently around snow patches, at altitudes of c. 1000 to 2800 m. The rather leathery, dark green leaves are kidney-shaped and grow on long stalks (petioles) from the base of the plant. The nodding, delicately violet to violet-blue flowers, which have a deeply fringed, open bell-like form are clustered in groups of up to four on long stems (up to c. 15 cm) and appear between April/May and August, frequently emerging through the melting snow (as Ruskin observes). As the flowers fade the stems elongate, holding the dry, oblong fruits (capsules) in an upright position.

32. See Diary, June 8th, above.
33. Ruskin's page 9 of the *Flora*.
34. Soils with a pH higher than 7, as would have been the case on the limestone rocks about Lac Bénit.

'c. d. Gnaphalium dioicum'
Now *Antennaria dioica* (L.) Gaertn. (Mountain Everlasting or Cat's-foot; male plants)
Asteraceae (syn. Compositae; Daisy or Aster family)

This stocky species is a low-growing, evergreen or semi-evergreen, mat-forming perennial with slender, branched horizontal stems (stolons) which spread out and produce new rooted shoots. It normally grows in dry mountain grasslands on calcareous soils (pH higher than 7) at altitudes of c. 1000 to 3000 m. There are two types of leaf: those at the base of the plant form a rosette and are oval to spoon-shaped, while the upper ones are relatively narrow, pointed, and spirally arranged on woolly stems. The lower surfaces of all leaves are downy and whitish in appearance, but the upper surfaces are usually hairless and grey-green in colour.

Antennaria dioica forms its male and female flowers on different plants (the technical term is dioecious), hence the species name. The flower-heads, which appear from May to July, are composite, comprising numerous, very small, tightly packed flowers (florets) arranged on a flattened structure at the top of a stem. The heads occur in groups of two to eight on short stems and become stiff and papery as they mature, hence the colloquial name 'everlasting flowers'. Each head comprises only disc florets (equivalent to, for example, the yellow, central florets of a Daisy flower head), surrounded by papery bracts (leaf-like structures), the outer ones with woolly undersides, which superficially resemble ray florets (equivalent to the radiating, petal shaped white florets of a Daisy flower). The flower heads of Ruskin's specimens c. and d. are from male plants,[35] which tend to be smaller than the females, are borne on shorter stalks and have white, spreading bracts, giving an overall white appearance. In contrast, the flower heads of female plants[36] are larger than those of the male plants, with pink disc florets in the centre, surrounded by pink bracts.

This species is probably referred to as 'No. 2' in Ruskin's Diary entry for June 5th (see above), as follows: 'The white plant, No. 1 [MYANTHEMUM BIFOLIUM, q.v.], very far down in the woods; No. 2 higher up [Gnaphalium dioicum].' No further information about the two specimens is provided in either the Diary or the hand-written notes in the *Flora of Chamouni 1844*.

35. See also the specimens relating to Ruskin's page 12 of the *Flora*.

36. Not collected by Ruskin on June 5th, but see the specimens relating to Ruskin's page 33 of the *Flora*.

Transcript of Ruskin's page 4

GLOBULARIA CORDIFOLIA

It was growing in masses which made the ground purple all about the Lac Benit;[37] associated with the Soldanella. Their delicate purples opposed to the gold of the Potentilla,[38] and the common crowfoot.[39] But its colour is now entirely gone. It was I think, much fresher in appearance when I first took it out of its envelopes, and has lost what remained to it of colour by exposure to the light. What vestige of hue remains to it, is however its original one, it has not turned purple from red or blue, as is the case with many plants.

Transcript of verso of Ruskin's specimen page

GENTIANA ALPINA

I seldom gathered the gentians—finding it impossible to preserve their colour, but I found them this day growing in such extraordinary masses on the very top of the Montagne Benit[40] that I gathered one or two for memorials. I have given away all but this, and I might as well have parted with this too—for it has not only lost its colour,[41] but shrunk to one half its original size, and is now a very slander on itself and its companions.[42] I never afterwards in all my Alpine rambles, met with this gentian either in such perfection or profusion—and owing to the early season, the flowers having just burst, their colour was intense beyond imagination. As the season advances, the colour diminishes greatly in purity & vigour; Fine specimens are only to be found by climbing high, to spots where their blossoming has been late. Thus a zone of bright blue, followed by one of pale blue, ascends the mountain sides as the summer advances.

37. Lac Bénit, which lies at the foot of the majestic limestone cliff of the Massif du Bargy.
38. A *Potentilla* species (a Cinquefoil).
39. A *Ranunculus* species (a Buttercup).
40. Perhaps Ruskin means the Massif du Bargy (see footnote 37).
41. In fact, some traces of the original blue colour remain in the petals.
42. In his Diary entry for June 5th (see above), Ruskin identifies this plant as 'bell gentian'.

Editor's note on specimens

'GLOBULARIA CORDIFOLIA'

Globularia cordifolia L. (Heart-leaved Globe Daisy)
Plantaginaceae (Plantain family)

This delightful species grows in niches and hollows on calcareous rocks, screes, rocky soils and grasslands, at elevations up to c. 2800 metres. It is an evergreen, creeping perennial sub-shrub, up to 10 cm high, which forms dense mats bearing basal, dark, glossy-green, spoon- to heart-shaped leaves (hence the species and vernacular names). The flowers, on short stalks, appear from May to July/August and are very small, soft lavender-blue in colour and clustered in fluffy heads (c. 1.5–2 cm diameter), which resemble miniature powder-puffs.

'GENTIANA ALPINA'

Ruskin identifies this specimen as: *Gentiana alpina* Vill. (Alpine or Southern Gentian). However, it more closely resembles: *Gentiana clusii* Perr. & Songeon (Clusius's Gentian), Gentianaceae (Gentian family)

Gentiana alpina (Alpine Gentian), a low (c. 4–9 cm), somewhat tufted perennial, prefers acidic soils (pH lower than 7) under rough grassland and among rocks, to an altitude of c. 2600 m. The roughly elliptical, pointed leaves, rarely more than c. 2 cm in length (they are up to 3 cm long in Ruskin's specimen), form a basal rosette and are pale green and rather fleshy. The solitary flowers, which appear from May to August, are held on very short, leafless stems and are intense blue in colour, the fused petals forming a dramatic trumpet with five rounded lobes at the rim and adorned with olive green spots in the throat. Although only c. 4 cm long, they appear huge compared to the rest of this diminutive plant. The fused sepals (calyx) form five lanceolate lobes.

Gentiana clusii (Clusius's Gentian), in contrast to Alpine Gentian, prefers the alkaline soils of rough grassland and rocky places on limestone (as does *Globularia cordifolia*, collected by Ruskin nearby), to altitudes of c. 2800 m. It flowers from May to August, and in form is rather like a larger form of Alpine Gentian. The pale green, pointed leaves of the basal rosette tend to be a little narrower and longer (3–3.5 cm) than those of Alpine Gentian, however, and are described as hard and leathery rather than fleshy. The solitary flowers are usually carried on longer stems than those of Alpine Gentian, with one to two pairs of small stem leaves (as in Ruskin's specimen, although the specimen stem is quite short). The dramatic solitary, trumpet-shaped, 'His Master's Voice' flowers are a medium to vivid deep-blue in colour (just as described by Ruskin), sometimes plain or with a few pale spots in the throat and up to 6 cm long (c. 4 cm in Ruskin's specimen; note, however, that Ruskin asserted that his specimen had 'shrunk to half its original size'). The calyx usually has five short, triangular lobes (again as in Ruskin's specimen).

Ruskin's specimen cannot be identified with certainty for various reasons, including the following: because of its historical importance, it could not be dissected to reveal those parts that were hidden; it had shrunk during drying and most of the colour had been lost; and there were no replicate specimens to allow confirmation of the forms and sizes of the leaves and flowers. Nevertheless, because it was collected in a limestone area and because it seems to have slightly more morphological features in common with *G. clusii* than with *G. alpina*, it is concluded that it is a somewhat attenuated specimen of *Gentiana clusii*.

A specimen identified as *Gentiana alpina* and a specimen with many similarities to *G. clusii*, but identified as *Gentiana latifolia* ou *excisa*, appear in Payot's *Herbier des Alpes*, p. 30.

Transcript of Ruskin's page 5

The two specimens were gathered by my Mother, from the meadows under the Nant d'Arpenaz, while I was absent on the Mont du Reposoir.

Editor's note on specimen

Possibly *Orobanche* species (Broomrape)
Family: Orobanchaceae

The specimens collected when Ruskin was elsewhere have disappeared, leaving only ethereal sepia images of the plants and the paper strips that once secured them. The images suggest, however, that the plants were c. 17–19 cm tall, that the stems had bulbous bases, and that the leaves were rudimentary, narrow, pointed scale-leaves, scattered up the stem, with no evidence of full sized leaves having been present. The flower heads were at the tip of each stem and appear to have comprised a short spike of between seven and ten relatively elongate (c. 1.5–2 cm), tubular, lipped flowers.

These features strongly suggest that the specimens were of a species of *Orobanche* (Broomrape), a genus of visually striking—almost ghostly—plants that lack the green pigment chlorophyll, and thus appear pale yellow, sometimes with a reddish tinge. Since chlorophyll is essential for the synthesis of the carbohydrates that all plants require for growth and metabolism, Broomrapes acquire this nutrient by parasitizing other species. Thus the seeds remain dormant in the soil until stimulated to germinate by chemicals released from the roots of potential host plants. As the seedling emerges, it produces a root-like structure which attaches itself to the roots of nearby host plants, taking soluble nutrients, especially sugars, from them. Most Broomrapes are relatively host-specific but without knowing the species it is not possible to say what the host of Mrs Ruskin's *Orabanche* might have been.

It might be argued that the images also appear to suggest the flowers of *Neottia nidus-avis* (L.) Rich. (Birds-nest Orchid), but this species normally grows in beechwoods, not meadows as recorded for these specimens.

It would seem that Margaret Ruskin had an eye for dramatic, unusual flowers, and in dismissing her specimens so perfunctorily her son John missed the opportunity of studying a member of a most fascinating and unusual group of plants.

*John Ruskin, Fringed Gentian (*Gentianopsis ciliata *(L.)), pencil and watercolour. Inscribed 'Les Rousses. Breakfast time 4th Sept 82. JR' and again 'see page 34 diary', this drawing was made at a much later date, and later in the year than the unidentified Gentians in the text, but it emphasizes Ruskin's love of both the form and colour of Gentians (see verso of Ruskin's page 4 of the* Flora of Chamouni 1844*)*

COLLECTING DAY TWO: JUNE 8TH, 1844

From the Diary: June 6th–8th, 1844

By June 6th the Ruskin's had reached Chamonix, and later in the day John Ruskin wrote: 'A lovely day, to light me to my own valley.' Soon after their arrival, Ruskin père provided his son with a new guide, Joseph Couttet;[1] the guide for their earlier visits, Michel Devouassoud,[2] had succumbed to a severe chill, and despite dosing himself with absinthe was close to death. Couttet, who was brought out of retirement by special dispensation, accompanied John Ruskin on the following day, June 7th, and for the remainder of the 1844 visit. This most knowledgeable, trustworthy and above all good-humoured man, was destined to play a significant role in Ruskin's plant collecting expeditions in the Chamonix valley and even contributed specimens of his own, as will become apparent. By 5.15 pm on the 6th Ruskin was in an optimistic frame of mind, even suggesting that the ennui of the early part of the day had lifted and all seemed set fair for the morrow; but it was not to be.

Couttet, or 'Coutet', as Ruskin referred to him in his collecting notes, duly arrived on the morning of the 7th, and he and Ruskin, together with the latter's manservant, George[3] Hobbs, who was no doubt deputed to carry Ruskin's somewhat bulky plant collecting vasculum[4] and probably drawing equipment as well, set out together for the base of the Buet, Ruskin on muleback and Couttet and George on foot. The clear morning gave way to 'a day of tremendous heat', however, Ruskin became ill ('Bilious chiefly!!') while ascending by the Tête Noir, towards the Buet, and as a result no plant specimens were acquired.

June 8th was more rewarding from a botanical point of view. Having rested until seven o'clock, the Ruskin party was on the road by 8.30 am, first by charabanc[5] to the mill half way to Argentière, where John was able to show his father and mother the

1. See Stephen Wildman, 'Ruskin and Chamonix', above.
2. *Works*, vol. XXXV, *Præterita*, pp. 328–329.
3. Actually John, but called George to distinguish him from both John Ruskin and his father.
4. See David S. Ingram, 'Alpine Botany and the *Flora of Chamouni 1844*', above.
5. See footnote 42, p. 51.

avalanche fallen from the Aiguille Verte. After this Ruskin, on a mule, together with Couttet and George, presumably on foot, climbed a steep path on the left side of the Flegère. On reaching the top they left the mule behind and 'had a staggering climb up long, slippery, dry, burnt grass', which 'took the shine' out of Ruskin before they had reached a thousand feet. 'After that, however, all went well: solid rock and moss, easy climbing to about eight hundred feet above the Flegere, where we found a little hollow in the rocks,[6] with a bed of Alpine roses,[7] on which we dined a l'antique, recumbent, with a rock, apparently on purpose for us, cleft, holding a little pool of cold snow water, clean by filtration through the moss, breast high, just below.' After 'dinner' they continued climbing, reaching 'a thousand feet higher', where they built a pyramid of stones before descending to the Flegère, crossing on the way 'some enormous fields of snow. … The view at this time was very noble—the Mont Blanc, rising above the near snow fields'. Ruskin did not arrive back at his lodgings until 9 pm and after arranging his 'flowers', sat down to write up his Diary for the day, declaring that: 'I enjoyed the whole day in my old way, and walked home all the way from the Flegere, level and all, with a child's springiness of mind and step.'

Editor's botanical note

Botanically, June 8th was very successful, with eight species being collected for the *Flora*. One duplicate specimen similar to a *Potentilla* species collected on June 5th, and an additional *Alchemilla* of unremembered provenance, were also added to the haul later, during mounting. The specimens actually collected on the 8th suggest that Ruskin probably encountered a range of habitats on the coarsely foliate and banded metamorphic gneiss and other rocks of the area, including acidic through to alkaline soils or rocky substrates, from the snowline margin down to sub-alpine and lower grasslands.

6. See also Transcription of Ruskin's page 8 of the *Flora*.
7. Presumably Alpenrose (either *Rhododendron ferrugineum* L. or *R. hirsutum* L., depending on whether the soil was acidic (below pH 7) or alkaline (above pH 7), respectively.

Transcript of Ruskin's page 6

ORNITHOGALUM FISTULOSUM.

All the specimens were gathered on the 8th of June, while ascending the steep pasturages above the pines,[8] which lie between the Flègere and the northernmost of the aiguilles Rouges. It is an abundant flower in all such situations.

Editor's note on specimens

'ORNITHOGALUM FISTULOSUM'
Now *Gagea bohemica* (Zauschn.) Schult. & Schult. f.
(Early Star of Bethlehem or Bohemian Gagea)
Liliaceae (Lily family)

Bohemian Gagea, a delightful spring flower that usually grows in dry stony and rocky places at altitudes between c. 400 and 2300 m, is a variable, relatively low and slender, hairless, bulb-forming perennial. There are usually two basal leaves, which are narrow and rush-like, often wavy or sometimes even coiled. Above these are two stem leaves that are closely alternate and overlapping, lanceolate, with somewhat pointed tips. The 1 to 3 bright yellow flowers per plant, which appear from February to May, are c. 10–20 mm diameter, with pointed petal-like tepals (petal-like structures which function as both sepals and petals).

Ruskin's specimens do not fit this description in all respects: the plants are c. 10 cm tall; no basal leaves are present (but may not have been collected by Ruskin); the stems appear to be slightly hairy; the flowering plants were collected as late as June; and the tepals are round-ended with some divided at the tip. In some ways the plants resemble other *Gagea* species, notably *G. fragifera* (Vill.) E. Bayer & G. López. This now rare species has hairy stems, somewhat round-ended tepals, flowers from May to July and is found on lime-poor soils in, for example, mountain meadows to c. 2500 m. Nevertheless, the balance of evidence leads me to the conclusion[9] that the specimens are indeed of *Gagea bohemica*.

A specimen identified as *Ornithogalum fistulosum* appears in Payot's *Herbier des Alpes*, p. 41.

8. The word pines is probably used here as a generic term for various unidentified conifers, all members of the Pine family (Pinaceae), but not necessarily members of the genus *Pinus*.

9. A conclusion later confirmed by Jim Jermyn (Alpine specialist); personal communication.

Transcript of Ruskin's page 7

GEUM MONTANUM.

Benoite de Montagne.[10]

Gathered the same day [June 8th], and abundant everywhere, but I think in a zone somewhat above the region of the preceding plant.

Editor's note on specimens

'GEUM MONTANUM'

Geum montanum L. (Alpine Avens)

Rosaceae (Rose family)

Alpine Avens is a bold, short, hairy perennial which grows in rocky, nutrient-poor mountain pastures, from c. 1500 to 2800 m. The compound leaves, most of which form a basal rosette, are coarsely pinnate (leaflets borne, feather-like, on either side of a central stalk), the leaflets being toothed, sometimes deeply so, with the terminal leaflet being very much larger than the others. The leaves on the flower stalks, however, are very small, lanceolate in form and deeply toothed. The striking, solitary flowers, which form from June to August, are bright golden yellow, usually with six petals, these often notched at the apex. The long green pointed sepals alternate with the petals and are clearly visible between them when viewed from above. The small, single-seeded fruits (not included with the specimens) have feathery tips and are dispersed by wind.

A specimen identified as a related species, *Geum reptans*, appears in Payot's *Herbier des Alpes*, p. 16. This species is somewhat similar in appearance to *G. montanum*, but the terminal leaflets are smaller, the petals lack a central notch, the habitat is restricted to gravelly, unstable substrates, and the plants spread by means of slender, horizontal stems or runners (absent from Ruskin's specimens), hence its colloquial name, Creeping Avens. I conclude, therefore, that Ruskin's specimens are indeed of *Geum montanum*.[11]

10. = Mountain avens.

11. A conclusion later confirmed by Jim Jermyn, personal communication.

Transcript of Ruskin's page 8

Gathered the same day [June 8th], merely as a reminiscence—(for the plant is everywhere abundant in the higher hills) of the place where we rested[12]—lying plunged over head and ears in masses of the Alpine rose,[13] under a grey bit of rock, which kept off the sun, and from a cleft in which tinkled a rivulet which we stopped with moss into a little crystal pool, and drank like horses; and then slept to the singing of it; the vision of the Mont Blanc, a great white throne, floating before ones eyes as they closed, entangled among the burning purples of the rhododendron [Alpenrose; see footnote 13].

Editor's note on specimens

These incomplete specimens, with one missing and the remaining three re-secured with thread instead of a paper strip, are probably of what is now called:

Kalmia procumbens (L.) Gift, Kron & P.F. Stevens ex Galasso, Banfi & F. Conti
(syn. *Loiseleuria procumbens* L. Loisel; Alpine, Creeping or Mountain Azalea)
Ericaceae (Heath family)

This diminutive azalea is a prostrate, wiry-stemmed, creeping and trailing evergreen sub-shrub that occurs in a range of relatively acid[14] mountain habitats at altitudes of between c. 1500 and 3000 m. The dark, shiny leaves, which occur in opposing pairs up the stem, are small, oval and hairless, with the edges rolled under. The very small, pink bell-like flowers with darker pink sepals grow in clusters at the tips of the branches and appear from May to July.

12. See also Diary entry for June 8th, 1844, above.
13. Probably *Rhododendron ferrugineum* L. (Alpenrose); Ericaceae (Heath family). These are evergreen shrubs, up to 1 m. high, with relatively small leaves that are shiny green above and covered with characteristic rusty red scales beneath. The bell-shaped flowers, which appear from May to July/August, vary in colour from pale to deep reddish-pink. This species prefers acidic mountain habitats, as does the *Loiseleuria procumbens* which Ruskin collected nearby on 'The same day' (see below). In contrast, *Rhododendron hirsutum* L. (Hairy Alpenrose), closely related to and sometimes confused with *R. ferrugineum*, prefers calcareous habitats.
14. The plant is widespread and even in limestone (alkaline) areas may be found growing on soils comprised of its own accumulated debris, where the pH may well be below 7; Jim Jermyn, personal communication.

Transcript of Ruskin's page 9

RANUNCULUS PYRENAEUS.

Very frequent in the Alpine pastures at four or five thousand feet above the sea. It should have come before the preceding plant, having been gathered a little lower with the Geum Montanum, p. 7.[15] The Soldanella[16] I found at the same time growing in considerable abundance, and larger than I ever saw it elsewhere.

.

Editor's note on specimens

'RANUNCULUS PYRENAEUS'

Upper specimens:

Ranunculus pyrenaeus L. (Pyrenean Buttercup) Ranunculaceae (Buttercup family)

Or, possibly:

Ranunculus kuepferi Greuter & Burdet (Kuepfer's Buttercup) Ranunculaceae (Buttercup family)

Pyrenean Buttercup is a very variable species. Although indigenous to the Pyrenees, it also occurs in spectacular snow-white drifts in damp grassland in the Alps, usually on alkaline soils, at altitudes of between c. 1500 and 3000 m. It is described as a short, largely hairless perennial with grass-like, narrow, linear or lanceolate, smooth-edged leaves without stalks (petioles) and with parallel veins. The flowers, which often begin to appear as the snow melts, are formed from spring into early summer. They are solitary, or sometimes in pairs or threes, at the ends of slender stalks. The five petals, with rounded tips, are pure white in colour, as are the sepals, and the whole flower is c. 1–2 cm diameter. There are numerous golden stamens at the base of the petals surrounding a cluster of ovules, which mature to become small, hard, single-seeded fruits called achenes, borne on a domed structure (the receptacle).

However, Kuepfer's Buttercup is also an Alpine species and is very similar in appearance to *Ranunculus pyrenaeus*. It is also very variable and forms large drifts, but unlike *R. pyrenaeus* grows on damp, acidic grassland soils. The flowers, which are formed from late spring to summer, are usually solitary (as with the specimens) and are slightly larger than those of *R. pyrenaeus*, being up to c. 3 cm diameter, with petals that may be slightly wavy or frilly.

Ruskin's pressed specimens are both rather dwarfed, but might fit with either of the above descriptions, especially given that both species are very variable. However, since Ruskin notes that *Soldanella* [*alpina*], a species that usually prefers calcareous (i.e. alkaline) soils, were 'growing in considerable abundance' nearby (and evidently collected a specimen at the same time: see the lower specimen on the same page), I conclude that the *Ranunculus* specimens here are of the species *R. pyrenaeus*.[17]

Lower specimen:

Soldanella alpina L. (Alpine Snowbell) Primulaceae (Primrose family)

For further details see Ruskin's page 3 et seq.

15. See Ruskin's page 7, this edition of the *Flora of Chamouni*.
16. The specimen below *Ranunculus pyrenaeus*; see Ruskin's page 3 et seq., this edition of the *Flora of Chamouni*.
17. A conclusion supported by Jim Jermyn; personal communication.

Transcript of Ruskin's page 10

These are exceedingly bad specimens of the flower, having been gathered above its proper zone; from the very highest rock I reached that day; near the foot of the aiguilles rouges; and at about the same elevation as the top of the Breven; I could not get further; for the rocks beyond were inaccessible except inch by inch; and our time was nearly exhausted. The specimens of the orchis nigra below (vide p.)[18] were gathered coming down from the wet meadow—the last bit of sloping ground in descending from the Flegere.

.

Editor's note on specimens

Upper specimens:
Primula integrifolia L. (Entire-leaved Primrose)[19]
Primulaceae (Primrose family)

Both of the miserable specimens here are, as Ruskin admits, 'exceedingly bad', having been gathered with great difficulty well above their normal elevation. Moreover, both specimens are poorly pressed and that on the right is incomplete. Nevertheless, they provide sufficient evidence for them to be identified as *Primula integrifolia* (Entire-leaved Primrose).

This small and lovely perennial prefers acid, stony or rocky, sparsely grassy substrates at altitudes between c. 1900 and 2700 m, often in snow hollows or by snow patches. The slightly sticky leaves form a basal rosette and are oval to lanceolate, virtually stalkless, with hairy and, as the species name suggests, smooth (un-toothed) margins. The flowers are large relative to the leaves and are borne in groups of 1 to 3 on (often reddish-stemmed) branched flattish flower heads (umbels). Each has five, deeply notched petals, which vary in colour from deep rose-purple through to a delicate pale pink or lilac.

One Primula specimen, identified as *Primula viscosa*, a synonym of *P. hirsuta* All. (Viscous or Hirsute Primrose), appears in Payot's *Herbier des Alpes*, p. 27.

Lower specimens:
The name 'orchis nigra', which Ruskin refers to in his collecting notes is now:
Gymnadenia rhellicani (Teppner & E. Klein) Teppner & E. Klein (Black Vanilla Orchid).
However, the specimens are probably of:
Gymnadenia conopsea (L.) R.Br. (Fragrant Orchid)
Orchidaceae (Orchid family)

Gymnadenia rhellicani has dark red to almost 'gothic' black flower heads and usually grows on slightly calcareous soils in mountain meadows and grasslands at altitudes of c. 1000 to 3000 m. However, the flower heads of Ruskin's specimens here are very much narrower than the wide, somewhat pyramidal form normally associated with this species. It has been suggested,[20]

18. No number given, but probably Ruskin's page 27 of this edition of the *Flora of Chamouni*.

19. Identified by Jim Jermyn; personal communication.

20. Jim Jermyn; personal communication.

therefore, that these specimens may in fact be of immature plants of Fragrant Orchid (*Gymnadenia conopsea*).

This fragrant species, vanilla-scented like Black Vanilla Orchid, prefers calcareous soils in meadows or at the margins of woodlands, at altitudes of up to c. 2800 m. The leaves are lanceolate, without spots, and the mature flower spikes, which develop from May/June until August, elongate to become almost cylindrical in form. These are densely packed with flowers, which range in colour from very dark to pale pink, with hooded petals, a short, three-lobed lip and a long, slender spur.

Gymnadenia rhellicani (Black Vanilla Orchid) is described in greater detail on Ruskin's page 27 of this edition of the *Flora*.

Transcript of Ruskin's page 11

POTENTILLA AUREA

I imagine these two specimens, a. and b. to be both the same flower, though the Blossom in one is so much more developed, at the expense of the strength of the plant. This smaller one, b, was gathered on the Mont. du Reposoir, June 5th, and was there associated with the leaves c, both plants growing in great quantities high and low. I cannot find the name of the plant c. and I have lost the ['to' crossed out] name of the locality from which I obtained the other potentilla, a.

Transcript of the verso of Ruskin's page 11

ALCHEMILLA VULGARIS. .C.

Editor's note on specimens

'POTENTILLA AUREA'

Upper specimens, labelled 'a' and 'b'.

These specimens are possibly of: *Potentilla aurea*, L. (Golden Cinquefoil), as suggested by Ruskin, but are more probably of:

Potentilla crantzii (Crantz) Beck ex Fritsch (Alpine Cinquefoil)

Rosaceae (Rose family)

Potentilla aurea is a low, mat-forming Alpine perennial which grows on non-calcareous/acid grassland soils at altitudes between c. 1300 and 3000 m. The flowers, which form from June to September, have bright golden yellow, overlapping, shallow-notched petals (not overlapping on the specimens), usually with an orangey-yellow spot at the base (cannot be detected on the dried specimens). The 'stipules' (small, leaf-like structures) at the base of each leaf and flower stalk are lanceolate. The compound leaves, however, are palmate, toothed, the end tooth of each leaflet being smaller than the others (not smaller on the specimens) and edged with silky hairs (not apparent on the specimens).

Potentilla crantzii grows in dry, calcareous Alpine grasslands at altitudes between c. 1300 and 3000 m. In this and many other respects it is similar to *P. aurea*, but the flowers, which form from May to September, have petals that do not overlap (as on the specimens) and lack an orange base. The leaves, on hairy stems (as on the specimens) are palmately lobed, toothed, with the end tooth equal in size to the others (as on the specimens), and are covered with flattened-down hairs on the upper surfaces (not visible on the specimens) and erect hairs on the lower surfaces (probably so on the specimens).

It is concluded that the specimens are probably of *Potentilla crantzii*, which is consistent with the observation that the specimen 'b' was collected on June 5th, when Ruskin was visiting a limestone location and collected other species that grow on calcareous soils.

Note that four other *Potentilla* species appear in Payot's *Herbier des Alpes*, pp. 15 and 16. One, on p. 15, is identified as *Potentilla salisburgensis*, a synonym of *Potentilla crantzii*.

COLLECTING DAY TWO: RUSKIN'S FLORA PAGE 11

'ALCHEMILLA VULGARIS'

Alchemilla vulgaris L. (Lady's Mantle)

(Syn. *A. acutiloba* Opiz)

Rosaceae (Rose family)

This familiar garden species grows wild in rich, moist grasslands to altitudes of c. 2800 m. The small, modest flowers (not present on the specimen) of this very variable, apomictic species[21] appear in June and July. They are borne in loose heads, lack petals, are soft greenish-yellow in colour, and rather inconspicuous. The leaves, often hairy, are roundish in overall form and divided to about half way to the base into seven to eleven tooth-edged lobes.

21. Produces seeds asexually.

COLLECTING DAY THREE: JUNE 9TH, 1844

From the Diary: June 9th 1844

The day, a Sunday, was 'intensely warm' and Ruskin sat in his room all morning '<u>reading divinity</u>', and 'watching the shadows of the pines[1] on the green sward of the Tapia'. He went out for a 'saunter' in the old meadows by the river before 'dinner', but it was 'too hot for enjoyment'. After dinner, he went for a 'better' ramble by himself, towards the Pèlerins, but it was still hot in the wood and Mont Blanc was under cloud. As he returned, after sitting reading a tract under a 'vast rock, rich with alpine rose', his botanist-artist's eye spotted a large, yellow anemone among blue gentian, on the north east of the Tête Noire pass, which he noted was 'very valuable as colour'.

Editor's botanical note

For Ruskin, Sundays would have been days of rest and spiritual contemplation, as he describes, and not plant collecting days. Nevertheless, he clearly could not help observing the plant life on June 9th and evidently allowed himself to collect three, admittedly relatively low altitude, but nevertheless interesting species for the *Flora*, the last of which was secured only after considerable exertion.

1. Presumably meaning various conifers in general.

Transcript of Ruskin's page 12

Moerhingia [sic][2]. G. 223.

I found this plant [upper specimens] frequently in damp places among the deep wood mosses; but these specimens were gathered between the Prieure and the source of the Arverom [sic],[3] in our quiet walk on the morning of Sunday, June 9th, just at the place where the bridge over the brook leads to the shattered block of granite & unfinished obelisk. The gnaphalia, below, were from the steep bank near the Cascade des Pelerins & were gathered the same evening—in my solitary walk. I took them for a rarer flower and had a scramble for them, swinging myself up by the pines; and there was a noble glimpse of stormy light down the valley at the time, which I should have lost, had it not been for the chase of the flower; which gave me command of it from a higher point.

Editor's note on specimens

Upper specimens
Possibly *Moehringia muscosa* L. (Mossy Sandwort)
Caryophyllaceae (Pink family)

'*Moerhingia*. G. 223' is clearly a reference to a genus colloquially described as 'Sandworts' (Pink family; Caryophyllaceae). The G. 223 is probably a reference to a specific illustration or page number in a botanical source describing a member of this genus. That the plant name and reference were written only lightly by Ruskin, or were partially erased after writing, suggests that he may have had second thoughts concerning his initial identification. This could have been because, although the specimens do resemble *Moehringia muscosa* L. (Mossy Sandwort), this species normally has only four sepals and petals, whereas Ruskin's specimens have five. Nevertheless, I tentatively suggest that Ruskin's identification may be correct[4], although other possibilities are noted below.

Moehringia muscosa, described by *Flora Europaea* as being 'very variable', is a perennial, loose tuft-forming species which prefers shaded rocky and mossy habitats, at altitudes from c. 1000–2000 m, as Ruskin suggests. The fragile flowering stems, c. 5–25 cm long, lack hairs, have paired, linear leaves and may bear one, but usually more (up to 6), very small, white flowers at the tips of the stems (Ruskin's specimens have only one flower per stem). These, as *Flora Europaea* also states, 'usually' have four petals, but as is sometimes the case with variable species, may on occasion possibly throw up variants that have five. Although the specimens of *M. muscosa* in the Herbarium of the Royal Botanic Garden Edinburgh all appear to have four petals, in general habit and form they do closely resemble Ruskin's specimens. Unfortunately, however, fruit capsules, which might have helped with the identification, were not present on Ruskin's specimens.

It is also noted by *Flora Europaea* that *M. muscosa* hybridizes frequently with *M. bavarica* (L.) Gren. (Narrow Leaved Sandwort) and *M. ciliata* (Scop.) Dalla Torre (Creeping Sandwort), both of

2. Moerhingia, very faint.
3. Presumably Arveyron.
4. I thank Dr Alan Elliott of the Royal Botanic Garden Edinburgh for his expert advice concerning this suggestion.

which have flowers with five petals. The hybrids usually have a mixture of four and five petalled flowers. Both species, however, unlike *M. muscosa*, are normally found on limestone substrates, although it is possible that a hybrid might not be restricted in this way. Alternatively, the spot where Ruskin says he collected the specimen may have been on calcareous strata exposed by river action.

Finally, *Flora Europaea* and Grey-Wilson & Blamey (1995) list another species, *Moehringia glaucovirens* Bertol., which is described as: having five-petalled flowers, usually solitary and smaller than the flowers of *M. muscosa*, and which may be found on shady limestone rocks in northern Italy. It is just possible, however, that *M. glaucovirens* did, in 1844, occur at the spot that Ruskin describes, perhaps as suggested above, on exposed calcareous strata, but has since died out there.

It may be concluded that the plants that Ruskin describes in his collecting notes as found 'frequently in damp places among the deep wood mosses', may indeed have been *M. muscosa*, which would have looked very similar to the specimen from 'between the Prieure and the source of the Arverom [sic]' included in the *Flora*, and may, therefore, easily have been assumed to be the same species. Alternatively, the specimen in the *Flora of Chamouni* may be a chance, five-petalled variant or hybrid of *M. muscosa*, or perhaps even a clump of *M. glaucovirens*.

The two lower specimens:
Antennaria dioica (L.) Gaertn. (syn. *Gnaphalium dioicum*; Mountain Everlasting, Cat's Foot; male plants)
Asteraceae (syn. Compositae; Daisy or Aster family)

For further details of these hard-won specimens of a species which, Ruskin was disappointed to discover, were not 'rarer', please refer to the notes to Collecting Day One (p. 60) and Collecting Day Nine (p. 113).

Transcript of Ruskin's page 13

SAXIFRAGA CUNEIFOLIA

This saxifrage,[5] common every where, was growing in great luxuriance (June 9th) in the lower woods of the Pelerins, whence I gathered these in my evening walk. But all the form of the flower is now lost.

Editor's note on specimens

'SAXIFRAGA CUNEIFOLIA'

Saxifraga[5] *cuneifolia* L. (Shield- or Spoon-leaved Saxifrage)

Saxifragaceae (Saxifrage family)

Shield-leaved Saxifrage, a perennial that spreads by means of specialized stems (stolons) that grow outwards, above or below ground to form new plants, prefers acid soils on moist, shaded and rocky ground in shrubby or woodland areas at altitudes of c. 600 to 2000 m. The characteristically shield-shaped, smooth, fleshy leaves normally have very shallow, blunt teeth[6] and form a basal rosette. The slender and leafless flower stem is c. 10–25 cm tall and bears small, delicate flowers with white petals and reflexed sepals in June and July. As Ruskin laments, those on the specimens have completely lost their elegant form during pressing and drying.

A specimen identified as *Saxifraga cuneifolia* appears in Payot's *Herbier des Alpes*, p. 14.

5. In *Proserpina* (*Works*, Vol. XXV, p. 315) Ruskin re-names this genus *Francesca*.
6. The leaf teeth on Ruskin's specimens seem almost as pronounced as those of *Saxifraga umbrosa* L., but the plants lack the characteristic hairiness and stem-leaves of this species.

COLLECTING DAY FOUR: JUNE 10TH, 1844

From the Diary: June 10th, 1844

Writing at '10 o'clock night', and after describing the sky at dawn, Ruskin notes that 'It turned out a day of awful heat.' He rode to the Tête Noire, but the heat fevered him and he could enjoy nothing. After 'dining' under a 'white, far falling fountain, bordered by green cushiony moss', he scrambled to the junction of the two rivers, then slept under a lime in a dell. Finally, he walked nearly all the way home, 'but headachy'.

Editor's botanical note

Interestingly, despite the heat, Ruskin did manage to acquire seven plant species for the *Flora of Chamouni* during June 10th, two contributed by Joseph Couttet, and most from moist or wet habitats.

Transcript of Ruskin's page 14

SAXIFRAGA ROTUNDIFOLIA

These plants I found June 10th, just under the gallery of the Tete Noire, growing among stones by a meadow streamlet. I never met with the flower in such perfection again.

Editor's note on specimens

'SAXIFRAGA ROTUNDIFOLIA'
Saxifraga[1] *rotundifolia* L. (Round-leaved Saxifrage)
Saxifragaceae (Saxifrage family)

This very large and variable Saxifrage is a perennial that prefers damp to shaded, sometimes rocky or wooded Alpine soils, often near springs or streams, as in this case, to an altitude of c. 2500 m. The leaves, which are usually slightly hairy, are round to kidney-shaped overall, but with variable indentations ranging from scalloped to deeply and sharply toothed, as with Ruskin's specimens. The tall, branched flower stems, which may be from 10 to 60 cm high, bear loose inflorescences from June to September. The individual flowers are relatively large, with white petals that have distinctive small red and yellow spots on the upper surfaces.

Although Ruskin collected only the tops of the plants, which, as he notes, were growing to 'perfection', the specimens nevertheless give a sense of the large size of this species.

1. See footnote 5, p. 78

Transcript of Ruskin's page 15

PRIMULA FARINOSA

Below the saxifrage places, we came to some steep meadows, sloping to the brink of the eau noire, where I found this primula, and I imagine the grasses below—Coutet was always gathering grasses; as if I had wanted a hay rick instead of a Flora, and these are probably his selection.

Editor's note on specimens

Upper specimens:
Primula farinosa L. (Bird's-eye Primrose)
Primulaceae (Primrose family)

This charming, small and variable perennial prefers acid soils on wet, marshy or peaty grasslands, at altitudes between c. 800 to 2900 m. The leaves are spoon-shaped to oval, with finely and unevenly toothed margins and with the lower surfaces covered with a white, mealy, but not unattractive deposit. The large, flattish flower heads (umbels) of up to eight flowers on short stalks, all arising from the tip of a long (up to c. 25 cm) stem, form from May to late July. The golden-eyed flowers vary in colour from delicate lilac to purple, and very occasionally pure white. The petals are fused at the base to form a corolla tube, which opens out at the top into five individual petals, each deeply divided into two lobes. The long, pointed sepals enclosing the corolla tube are covered, like the leaves, with a mealy white deposit.

Lower specimens:
Carex sp. (a Sedge)[2]
Cyperaceae (Sedge family)

These damaged specimens are not of grasses, as Ruskin suggests, but of grass-like sedges. Assuming that Ruskin's collecting notes are accurate, they were probably growing in the same place as the *Primula farinosa*, above, in a meadow sloping down to l'Eau Noire, where the soil was probably wet and somewhat acidic. The specimens are clearly of a relatively small, clump forming species, collected early in the flowering season when the young male flowers were fully developed, the upper spike being covered with prominent anthers, giving it a ragged appearance, while the female flowers had still not matured. The glumes—the pointed leaf-like structures which enclose the individual flowers—are dark brown to black, with lighter membranous edges. The overall appearance of the three spikes on each plant, the upper one relatively large and the lower two small and overlapping, is dark brown to black with lighter patterning.

The specimen plants are not sufficiently well developed for it to be possible to identify them with certainty, but to my eye they could be of *Carex nigra* (L.) Reichard (Common Sedge).

2. I thank Dr Phil Lusby of the Royal Botanic Garden Edinburgh for discussion of the identity of these specimens.

Transcript of Ruskin's page 16

These grasses [sic] were from the same meadows, but are common everywhere. [T]hey were intended to form part of a series of grasses; but I was not again among the low meadows, and got only hill plants.

Editor's note on specimens

None of these specimens is a true grass, although all are grass-like.

Left-hand specimen:
Eriophorum scheuzeri Hoppe (Scheuzer's Cotton-grass; see also Ruskin's page 17 of the *Flora*)
Cyperaceae (Sedge family)

This attractive sedge usually prefers wet meadows close to the water at altitudes between 2000 and 2800 m. The flowering heads, which form in early spring, are covered with the bright yellow anthers of the many tiny flowers, but later they become covered with the long white hairs attached to the fruits, as in this specimen. A related, single-headed *Eriophorum* species, *E. vaginatum* L. (Hare's-tail Cotton Grass), tends to be somewhat taller than *E. scheuzeri* and normally grows on very acidic, peat and boggy substrates.

Two centre specimens:
Bistorta vivipara (L.) Delarbre (Alpine Bistort) (syn. *Persicaria vivipara* (L.) Rons. Decre.)[3]
Polygonaceae (Dock and Sorrel family)

This low growing, somewhat grass-like plant prefers moist grassy and rocky meadows and other grassy places at altitudes between 1000 and 2500 m. The hairless leaves are oblong to narrow and lanceolate, tapering towards the base, usually with the margin rolled under. The very small flowers, which form at the upper ends of slender spikes from June to August, are a delicate pale pink to white in colour. The lower parts of the spikes carry, instead of flowers, small, brownish-purple bulbils, formed without pollination, and these are capable of growing into new plants when they fall to the ground (hence the species name *vivipera*).

Right-hand specimen:
Bistorta officinalis Delebr. (Bistort) (syn. *Polygonum bistorta* L.)
Polygonaceae (Dock and Sorrel family)

This common and slightly more robust, patch-forming close relative of Alpine Bistort grows in damp meadows and other moist grassy places at altitudes of 1000 to 2500 m. The plants arise from spreading, underground stems (rhizomes), and the hairless leaves are narrow-oval to elongate-triangular, with the leaf stalks (petioles) slightly winged towards the leaf blades (laminas). The small, closely packed, dark pink flowers form in wide, unbranched spikes (without bulbils) from late May to late August/September.

3. I thank Dr. Henry Noltie, Royal Botanic Garden Edinburgh, for his advice on the identity of this species.

Transcript of Ruskin's page 17

In returning, the same afternoon, June 10th, I walked as far as the tarn up the valley to the col de Berard. There I waited for Coutet, whom I had lost sight of among the rocks between the Chalets de Berard and the Tete noire road. The meadows underneath were always covered with the bell gentian[4] and globe Ranunculus[5]: the latter peculiarly fine, but I only gathered these grasses, for fear of crushing my primulas and saxifrages.[6]

Editor's note on specimens

Eriophorum scheuzeri Hoppe (Scheuzer's Cotton-grass)
Cyperaceae (Sedge family)

These fluffy-headed, delicate grass-like specimens are all of the same species, which is not a true grass but a member of the Sedge family. The centre specimen is in full flower, and those to its left and right are forming hairy fruits.

For a full description, see the Editor's note to Ruskin's page 16 of the *Flora*, opposite.

4. Presumably *Gentiana alpina* Vill. or *G. clusii* Perr. & Songeon; see Ruskin's page 4 of the *Flora*.
5. *Trollius europaeus* L.
6. See Ruskin's pages 14 and 15 of the *Flora*.

Transcript of Ruskin's page 18

When Josef[7] came up, the same afternoon, he brought me these junci[8] as if they were rare. I kept them therefore, though they seemed to me one of the most frequent meadow plants.

Editor's note on specimen

Luzula campestris (L.) DC.
(syn. *Juncus campestris* L.; Field Wood-rush)
Juncaceae (Rush family)

This elegant (10–30 cm.), widely distributed and very variable upland rush, which spreads by means of short, horizontal stems (runners), prefers dry meadows at altitudes of between c. 700 and 2600 m. The height of its slender, hairy stems varies from c. 10–30 cm. The leaves, also hairy, are linear and grass-like. The tightly packed clusters (glomerules) of small, slightly coarse, russet-black flowers are arranged in umbels of 2 to 6 at the tips of the stems. In the Alps the flower heads appear from June to August and during the early part of the season are covered with bright yellow, ephemeral stamens.

A somewhat similar rush, *Luzula sudetica* (Willd.) Schult., has distinctly reddish stems and upper leaves, prefers wet meadow and lake margin habitats and has a much narrower altitude range (c. 1700–2200 m) than *L. campestris*.

7. Joseph Couttet, Ruskin's local guide; see the Diary entry for June 7th, p. 65.

8. Rushes (members of the genus *Juncus*).

COLLECTING DAY FIVE: JUNE 11TH, 1844

From the Diary: June 11th, 1844

This was evidently a good day, which began with a soft shower that cleared at 10 am and left the hills 'glowing under the hot sun'. Ruskin set off with Couttet 'for a quiet saunter' through the wood towards the Arveyron arch where he found 'some new mosses'. Later they climbed over glacial moraine until they came to a very large block of Dru granite, which Couttet declared was the largest in the valley. After Hobbs, Ruskin's manservant, had recklessly climbed it and had to be helped down by Couttet 'with the aid of a large pine trunk', more hard climbing over debris brought them to some 'calcareous slates en place under the glacier'. Unfortunately, no more of the local geology can be gleaned from the manuscript Diary since half a page was cut out at some point and is now lost.

Editor's botanical note

This turned out to be a most interesting day, botanically, with six species being collected. Two were different, but related, moss species and three were very varied *Saxifraga* species. It is important that the Diary mentions that some of the metamorphic rocks were calcareous, for this emphasizes that alkaline substrates, in addition to acid and neutral ones, occur in the Chamonix valley, as the species mix for the day demonstrates. It should be noted, however, that the pH of soils in the Alps, as in most places, does not necessarily reflect the pH of the underlying rocks, for these may sometimes be overlain with boulder clays, muds, peats and other accumulated plant-derived materials with very different pH and other properties.

Transcript of Ruskin's page 19

These mosses were gathered June 11th, in walking to the Chapeau through the Arveron woods, together with a plant of the Pinguicula, vid P.30,[1] which was growing luxuriantly in the green bed of a little running stream. The moss is very frequent in the woods.

Editor's note on specimens

Polytrichum commune Hedw.[2]
(Common Haircap Moss, Great Golden Maidenhair Moss or Great Goldilocks Moss; male plants)[3]
Polytrichaceae (one of the many families of Mosses[4] [Musci]).

The Mosses are very ancient (evolutionarily 'primitive') land plants, which have complex life cycles and lack flowers in a conventional sense. They also lack a waterproof outer layer (cuticle), that in most plants prevents excessive water loss to the atmosphere, and require free water to enable the swimming male spores to fertilize the female egg cell during sexual reproduction. They are, therefore, usually restricted to damp, humid areas with a relatively high rainfall.

The tall, slender plants of *Polytrichum commune* preserved here were clearly pulled from a characteristically deep mat of tightly packed, tall slender shoots growing in the wet woodland habitat that Ruskin describes in his accompanying notes. Species of *Polytrichum* are dioecious, producing male and female reproductive structures on separate plants.[5] The specimens here are all single-stemmed, male plants. Each terminates in a male 'false-flower' (false-inflorescence), made up of small leaf-like structures (bracts), which enclose the male reproductive organs (antheridia). Following sexual reproduction a new male shoot grows out of the by then defunct male false-flower, and this in turn forms at its tip a new false-flower the following year. The result is a tiered structure made up of a series of male shoots, one upon another like a circus balancing act. Some of the plants here show such evidence of at least three years' growth.

1. See Ruskin's page 30 of the *Flora*.
2. Drs D. Long and P. Lusby of the Royal Botanic Garden Edinburgh are thanked for providing expert advice concerning the identity of this species.
3. See for these and other names for this species: Edwards, S. R., *English Names for British Bryophytes*, British Bryological Society Special Volume No. 5 (Wootton, Northampton; 4th edition, 2012).
4. Ruskin deals with mosses as a group in Chapter 1 of *Proserpina*, where his approach is characteristically idiosyncratic, but nevertheless fascinating and enjoyable (see my chapter 'Alpine Botany and the *Flora of Chamouni 1844*' in this volume).
5. As is the case with some flowering plants such as, for example, *Antennaria dioica* (L.) Gaertn.

Transcript of Ruskin's page 20

This moss was gathered in the same spot as the preceding, but it occurs in far greater abundance among the dry stones left by the great avalanche beneath the cascade des Pelerins, growing in deep & thick masses.

Editor's note on specimens

Polytrichastrum alpinum (Hedw.) G.L. Sm.[6]
(formerly *Polytrichum alpinum*; Alpine Haircap Moss;
female plants with capsules)
Polytrichaceae (one of the many families of Mosses [Musci])

See the Editor's note to Ruskin's page 19 of the *Flora* (opposite) for general information concerning mosses.

This moss species is closely related to *Polytrichum commune* (see Ruskin's page 19). However, in contrast to that species, it is characteristic of dry rocky places, as described in Ruskin's notes. Moreover, the dried leaves of the specimens are strongly appressed to the stems, a water-conserving strategy, and one plant is branched (rare in the Polytrichaceae), both of these features being characteristic of *Polytrichastrum alpinum*. Like *P. commune*, *Polytrichastrum alpinum* occurs as separate male and female plants (it is dioecious), the leafy specimens here all being female.

Emerging from the apex of each of the female plants is a young, leafless spore-producing plant, the sporophyte (2 in one case). Since moss sporophytes lack the green pigment chlorophyll and cannot, therefore, manufacture their own sugars for metabolism, they are parasitic on the female plants that formed them. At the apex of each sporophyte is a cylindrical capsule, covered by a hairy[7] cap (the calyptra). The sporophytes would have formed from the female egg cells following fertilization and the capsules would, when mature, have released dust-like spores (equivalent to seeds) to be dispersed to new habitats by the wind.

6. Drs D. Long and P. Lusby of the Royal Botanic Garden Edinburgh are thanked for their expert advice concerning the identity of this species.
7. Hence the vernacular and scientific species and family names. See W. T. Stearn, *Botanical Latin*, (David & Charles, Newton Abbot, 3rd edition, 1983).

Transcript of Ruskin's page 21

SAXIFRAGA COTYLEDON. L

Saxifraga aizòòn[8] Jaq.

Gathered 11th June, in my scramble with Coutet up the moraine of the Glacier des Bois, just under the Chapeau, about the place where the granite joins the Buet slate. This saxifrage was growing on the patches of soil which had lodged in the hollows of the moraine.

Editor's note on specimens

'SAXIFRAGA COTYLEDON. L'

Saxifraga cotyledon L. (Pyramidal Saxifrage)

Or, more probably:

Saxifraga paniculata Mill. (syn. *Saxifraga aizoon* Jacq.; Paniculate or Livelong Saxifrage)

Saxifragaceae (Saxifrage family)

That Ruskin gave these specimens two alternative scientific names suggests that he was uncertain of their correct identity, as am I.

Pyramidal Saxifrage grows in crevices in acid rocks, such as granites, at altitudes from c. 1500 to 2500 m. It has white flowers that appear from June to August. However, the very small size (only c. 4 cm) of the basal rosettes of leaves on the specimens (expected to be up to 15 cm), the deposits of lime on the leaves (expected to be lime-free), and the small size and simple structure of the panicles of flowers (expected to have very large panicles, branched low down and carrying up to 100 flowers[9]), all suggest that the specimens are not of *S. cotyledon*.

The alternative species name suggested by Ruskin, *Saxifraga aizoon* (now *Saxifraga paniculata*), is a very variable Saxifrage that produces white flowers with red spots on their upper surfaces, from May to August. This name is more likely to be the correct one since the species is described as: occurring in rocky places to c. 3000 m (some, but not all, authors suggest calcareous rocks, a possibility in the position described by Ruskin); having relatively small rosettes of leaves with lime encrustations; and in having simpler panicles of fewer flowers than *S. cotyledon* (this characteristic being accentuated in the specimens by the fact that the panicles are not mature and most of the flowers are still in bud).

8. Meaning similar to an ice plant (a member of the mainly subtropical and tropical family Aizoaceae).

9. See, for example, specimen 13 in Payot's *Herbier des Alpes*.

Transcript of Ruskin's page 22

SAXIFRAGA ASPERA

Saxifrage rude[10]

Just beneath the chapeau, and at the edge of the moraine of the Glacier des Bois there are some projecting rocks (granite) en place, which are mainly bare of vegetation and consequently exhibit, better than any that I know, the effect of the glacial action. On the tops of these, however, this, & the other hardy saxifrages grow in greater perfection than I ever found elsewhere.

Editor's note on specimens

'SAXIFRAGA ASPERA'

Saxifraga aspera L. (Rough Saxifrage)

Saxifragaceae (Saxifrage family)

This now rare species grows in acid soil in stony and rocky places or in rock crevices, as described by Ruskin in his notes, at altitudes from c. 1200 to 2500 m. The leaves, most of which have long, stiff hairs, do not form a basal rosette, although miniature rosettes of small leaves often form in the axils of the larger, lower leaves of the straggling, 5–20 cm, stems. The groups of from 2 to 6 dainty flowers are carried on a slender, branching stem and appear from June/July to August. Each flower has prominent anthers and five creamy-white petals shading to soft yellow at the base, with yellow or red/violet spots on the upper surface.

10. Meaning 'rough Saxifrage'.

Transcript of Ruskin's page 23

SAXIFRAGA MUSCOIDES

Saxifrage mousse[11]

From the same rocks[12] as the preceding species.

Transcript of verso of Ruskin's page 23

POLYGALA ALPINA

Polygale des Alpes.

This lovely plant grows luxuriantly on the same rocks as the preceding saxifrages. I do not think it is uncommon, but have not found it elsewhere. It was not on the barren parts of the rocks, but just where the rich wood vegetation began to appear.

Editor's note on specimens

'SAXIFRAGA MUSCOIDES'

Upper specimens:

The leaves of *Saxifraga muscoides* All. are lanceolate, without lobes, whereas the leaves of these specimens are divided into three straight sided, blunt ended lobes, features which suggest that the correct identity may be:

Saxifraga moschata Wulfen (Musky Saxifrage)

Saxifragaceae (Saxifrage family)

This common, variable species grows on stony ground or among rocks where there is some moisture, and is occasionally found on the edges of snowfields, to altitudes between c. 1200 to 3500+m. The leaves are usually divided into three straight sided, blunt ended lobes, as here, but may sometimes be un-lobed or five lobed, and form dense cushions at the base of the plant. The flowers, which appear from June/July to August, are usually a lovely pale yellow-green colour, but may occasionally be white, and are borne in branching flower heads (panicles) at the tips of stiff stems.

11. Meaning moss Saxifrage.

12. Projecting granite rocks at the edge of the moraine of the Glacier des Bois.

COLLECTING DAY FIVE: RUSKIN'S FLORA PAGE 23

'POLYGALA ALPINA'

Lower specimens:

The lack of a false basal rosette, the shape and disposition of the leaves (ovate, becoming longer and more lanceolate towards the apex; and alternate rather than opposite), and the form of the flower spike, all suggest that these specimens (one of which has been re-secured with twine) are not of *Polygala alpina* (DC.) Steud., as Ruskin's label implies, but are of:

Polygala alpestris Rchb. (Mountain Milkwort)

Polygalaceae (Milkwort family)

This, like so many Alpine plants, is a very variable species, especially given the diversity of habitat types and possibilities in the Alps. It does not usually grow on rocks, but in grassy places, as Ruskin's collecting notes suggest, usually at altitudes between c. 1000 and 2600 m. The leaves are as described above, on stems up to c. 15 cm high. The very pretty, light to dark blue, sometimes white flowers appear from May/June to July and are tightly grouped at the top of the stem, as in the specimens.

John Ruskin, drawing from 1844 diary notebook.
This depicts what Ruskin described in his diary entry for June 12th, 1844 as 'Mélèze', a French word meaning, loosely, any member of the Pine family (Pinaceae), but usually translated as Larch or Larix

COLLECTING DAY SIX: JUNE 12TH, 1844

From the Diary: June 12th, 1844

After a walk to the Pèlerins, in which he was stopped by rain, Ruskin sat down to investigate the mélèze, which he illustrates on the page facing the entry for June 6th of the manuscript diary for 1844 (reproduced opposite).[1]

Ruskin also provides a commentary on the illustrations, as follows:

> The filaments *c* are striated and about one half the size of figure; they are arranged on the bough so as to give the transverse section *a*, the circular part being uppermost, though for convenience I reversed in the drawing, the stem of course in the centre. The filaments at *x* are set on the bough at an angle of rather less than 45, while those at *y* and *z* are set on at rather more than 70. This causes the distance *nx*, although increased by the thickness of the stalk, to be materially less than the distance *ny* or *nz*. When the bough is seen from above (*b*) the filaments at *x*, of course, appear crossing and intersecting those at *y* and *z*. This gives variety and intricacy to the arrangement.
>
> The twigs branch irregularly from the main bough and from each other. The filamental foliage on any main bough does not reach much more than a foot from its extremity. Any side twig not more than a foot long will be partially clothed; if longer, it becomes bare and assumes the character of a main bough. The roots of the leaves are at first left visible on the bare bough, but not for above six inches; it then becomes smooth.
>
> The twigs show a marked tendency to leave the main bough in pairs; but those on one side (I cannot make out from my bough whether outer or inner—I presume the former) seem to flourish, and those on the other to grow thin and drop off (*d*). The thin ones run to a great length, so as to be unable to support themselves. The bough *f* is more than a third of an inch in diameter, but is not much longer than the bough *e*, which is the twelfth of an inch in diameter.

After next describing the scene while he is writing, Ruskin finally returns to the

1. MS4 held in the Whitehouse Collection, The Ruskin, Lancaster University.

morning, saying that he was diverted from his purpose of climbing the Montanvert by clouds, which he suggests are an indication of wet weather to come. Nevertheless, he says, they (presumably Ruskin, Couttet and Hobbs) got to the Cascade des Pèlerins 'over the ruins of a thousand pines [sic], splintered to ribbons by an avalanche during the previous March'. Finally he concludes that it was 'A good day.'

Editor's botanical note

The largest drawing accompanying the entry for June 12th appears in *Modern Painters*, reversed and on its side, as Plate 13, 'Study of Foliage'.[2] It is not referred to in the text, however, this omission being explained by a footnote to the 'List of Illustrations' on p. xvi of Vol. I,[3] describing it as one of those which 'have not before been published.'

The identity of the tree depicted on p. 92 is not stated, although Ruskin refers to it as 'Mélèze', a word usually translated as 'Larch' or '*Larix*', presumably European Larch (*Larix decidua* Mill.), a common deciduous conifer species in the Chamonix region. However, although not a large-scale drawing, the arrangement of the needles (leaves), to my eyes does not suggest Larch, where the needles are borne in discrete clusters along the branch. Two other species of conifer which are now common about Chamonix and do have the general form depicted in the drawing are: Silver Fir (*Abies alba* Mill.) and Norway Spruce (*Picea abies* (L.) H.Karst.), both, like *Larix decidua*, members of the family Pinacea. Of these two, and based only on my general impression of the flattened arrangement of the leaves and the detailed drawing of the single leaf labelled '*c*', with its clearly defined midrib, I tentatively suggest that the species depicted in the drawings is Silver Fir.

Ruskin drew and described another conifer specimen in his Diary entry for June 16th, 1844. I suggest, therefore, that the present note should be read in conjunction with Ruskin's 1844 diary entry for June 15th–17th and my Editor's botanical note for Collecting Day 8, this volume. Illustrated notes on the drawings are also included in Ingram & Wildman (2011).[4]

Trees, especially conifers, cannot easily be pressed and mounted in an album, hence, perhaps, the detailed drawings here and on p. 16 of the Diary. Four contrasting, dramatic and easily pressed species, which did not therefore need to be drawn, were in fact collected on June 12th and included in the *Flora*.

2. *Modern Painters I*, following p. 588 in vol. I, Section VI, 'Of Truth of Vegetation', *Works*, vol. III.
3. *Works*, vol. III, p. xvi.
4. Ingram, David [S.] & Wildman, S., *Ruskin's Flora: The Botanical Drawings of John Ruskin*, (Ruskin Library and Research Centre, Lancaster University, 2011), pp. 22–23.

Transcript of Ruskin's page 24

Gathered in the marshy ground under the cascade des Pelerins, as we returned from our stormy morning walk, June 12th.

Editor's note on specimens

Cardamine sp., perhaps a variant of *Cardamine pratensis* L. (Cuckoo Flower or Lady's Smock)
Brassicaceae (sometimes referred to as Cruciferae; Cress and Cabbage family)

At first sight these specimens suggest *Cardamine pratensis* L. (Lady's Smock or Cuckoo flower), a widespread and extremely variable species not confined to the Alps. It inhabits damp pastures, marshy ground and stream sides, to an altitude of c. 2600 m, just as Ruskin describes. Moreover, its flowers, with four petals, form between April and July, so by the time of Ruskin's visit both these and the long, narrow fruits (two chambered capsules which, when dry, split open from the base to release the seeds), would have been present on the plants in addition to the flowers, as in the specimens. However, most of the characteristic compound (pinnate) leaves, with 1–7 pairs of leaflets arranged, feather-like, along the main axis, plus a large terminal leaflet, usually grow as a basal rosette, a feature that these specimens lack, perhaps because Ruskin cut the stems above the rosette in order to accommodate the plants on a single page of his *Flora*.

Further examination of the specimens reveals another feature that is not consistent with what might be expected in *C. pratensis*, the presence of two types of leaves on the stem. *Cardamine pratensis* normally has on its stems only small compound leaves with narrow leaflets without notches, the larger compound leaves with wide, notched leaflets being restricted to the basal rosette. In contrast, the specimens carry both types of leaves on the stems, the smaller ones with narrow leaflets being uppermost and the larger ones with wider leaflets being lower down. It is possible, therefore, that these specimens are of a morphological variant of *Cardamine pratensis*. A variant that would be consistent with the form and habitat of the specimens here was in the past described as a different species, *Cardamine palustris* (Wimm. & Grab.) Peterm., a synonym of *Rorippa palustris* (L.) Besser. An excellent drawing of this variant, clearly showing both types of compound leaves on the stem and labelled as *Cardamine palustris*, is illustrated on p. 141 of Blamey & Grey-Wilson (2003).[5] If this identification is correct, the colour of the flowers on the specimens before drying would have been, as in *C. pratensis*, white, tinged with delicate pink to violet, and with equally delicate but slightly darker veining, a colour combination that would almost certainly have appealed to Ruskin's eye for beauty.

An alternative suggestion is that the specimens are of a different *Cardamine* species. One possibility is *C. crassifolia* Pourr., a rarer species than *C. pratensis*, usually found in wet mountain habitats, which often does not form a basal rosette. However, this species is not usually found in the high Alps. No other Alpine *Cardamine* species is consistent with both the form of the specimens and the location in which they were found, except perhaps *C. resedifolia* L., but in this species the lower leaves tend to be entire (not compound) or slightly lobed, and the upper ones, although very deeply lobed, are not actually compound.

It is concluded, therefore, that these specimens are indeed of a morphological variant of *Cardamine pratensis* L. (Cuckoo Flower or Lady's Smock).

5. Blamey, Marjorie & Grey-Wilson, Christopher, *Wild Flowers of Britain & Northern Europe*, (Weidenfeld & Nicolson, London, 2003).

Transcript of Ruskin's page 25

Both specimens[6] were gathered on June 12th, from the short—grassed—steep pastures under the Cascade des Pelerins. It is a common flower, but I never saw it of so vivid an orange as at this spot. Its colour was quite lost in drying.

Editor's note on specimens

Crepis aurea (L.) Cass. (Golden Hawk's-beard)
Asteraceae (syn. Compositae; Aster or Daisy family)

This spectacular species grows in moist, stony grasslands such as meadows and pastures, at altitudes of between c. 1800 to 2800 m. The leaves all form a basal rosette and are toothed, the dentation being variable, but in Ruskin's specimens deep and sharply angular. The composite flower heads (i.e. consisting of many small flowers clustered together on a single flat structure, the capitulum, at the tip of the stem) appear from June to August and are borne singly on long, slender stems. These are often covered, especially near the top, with long, dark hairs, which extend to and cover the small leafy bracts beneath the flower heads. The florets are all ray-florets, as in a dandelion head, and each has a blunt, serrated tip. All are of a soft, golden-orange colour, those on Ruskin's specimens clearly being of an unusually intense hue, sadly lost during drying and pressing. The elongate, hard dry fruits (achenes), terminate in a tuft of white silky hairs that, following ripening, may carry them on the wind.

6. The left hand specimen has been re-secured with twine.

Transcript of Ruskin's page 26

This plant, (a large shrub) grows luxuriantly in the clefts & shady hollows under the two small cascades east of the Casc. des Pelerins. I did not meet with it elsewhere.

Editor's note on specimens

Lonicera L. species, possibly *Lonicera nigra* L.
(Black-berried Honeysuckle)
Caprifoliaceae (Honeysuckle family)

This is clearly a fine, shrubby, non-climbing honeysuckle (*Lonicera* species). The paired, ovoid to slightly pointed, delicate-looking leaves with short petioles on straight stems, and the paired, asymmetric flowers on long stalks arising in the axils of the paired leaves are features consistent with the identification of this genus. The lack of fruits, damage to the open flowers and lack of flower colour on the specimens, however, makes precise identification of the species difficult. Nevertheless, on the basis of the leaf shape and overall form of the specimens, it is suggested that they may be of *Lonicera nigra* L. (Black-berried Honeysuckle), rather than *L. alpigena* L. or *L. xylosteum* L. This deciduous, twiggy sub-shrub, which can grow from 1–2 m high, may be found in wet woodland (especially beech and conifer) and in stony or grassy places to an altitude of c. 2500 m. Its leaves are elliptical to oblong and usually smooth. The paired, pale whitish-pink, two-lipped and delicately fragrant flowers are produced from May to July and, as the name suggests, round black (inedible) berries, not present on the specimens, are produced later in the season.

Transcript of Ruskin's page 27

ORCHIS NIGRA

Also from the meadows near the cascade des Pelerins June 12th. It is not unfrequent, & its vivid point of black is a striking feature of an Alpine pasture.

Editor's note on specimens

'ORCHIS NIGRA'
Ruskin may be referring to what is now named:
Gymnadenia rhellicani (Teppner & E. Klein) Teppner & E. Klein (Black Vanilla Orchid)
Orchidaceae (Orchid family)

The black, vanilla scented Orchids as a group are difficult to identify with precision. They are now rare species which usually prefer poor, often slightly calcareous, dry to slightly moist soils in stony mountain meadows and grasslands at altitudes between c. 1000 to 3000 m. The pointed, narrow leaves are lanceolate to linear, almost grass-like, without spots. Visible at the base of the plant in one of Ruskin's specimens is the tuberous root, hence the family name.[6] The large, oval to pyramidal, densely packed flower heads form from June to August. When fresh they appear a very dark, dramatic red/purple colour, almost black (as Ruskin's scientific name for the specimens implies), but sometimes a slightly less striking dark pink. When fully open they have a spreading form with a relatively long, triangular and upturned upper lip.

6. Orchis is a Classical Greek word meaning a testicle, hence the origin of the vernacular name of the group, the former genus name of this species and the current name of the family.

COLLECTING DAY SEVEN: JUNE 14TH, 1844

From the Diary: June 13th & 14th, 1844

Although Ruskin had rather an energetic day on June 13th, 'climbing the Montanvert beside the glacier ... a thorough scramble, clinging to branches and swinging from trunks of trees in a capital style', no specimens were collected for the *Flora*. He does, however, describe resting among Alpine roses (Alpenrose; *Rhododendron ferrugineum* or *R. hirsutum*, depending on the soil pH) during the afternoon, before making the botanically interesting comment that 'I was surprised by the number of bees and butterflies [all potential pollinators of Alpine plants] at the top of the Montanvert.'

He returns to the subject of insects in the entry for the following day, writing '... much astonished to-day by the number of flies high up on the snow above the Montanvert.' June 14th was another 'delicious day, though a mighty hard one'. After setting off very early, at a leisurely pace for the Montenvers, he packed a knapsack for Couttet and they climbed quietly and slowly up the steep but mossy and soft hillside for an hour, followed by 'stony work' to reach the ridge, which gave them a cloudless view of Chamonix and the Mont du Reposoir. Next, there followed a difficult descent to the snow, followed by a climb of an hour or more on steep, inclined planes of snow, interspersed with formidable loose rocks. Finally, they reached the base of the Petit Charmoz, with Ruskin being 'amazed' at the distortions of size and distance that one experiences when climbing in the Alps. Towards the end of the day they climbed for about 30 minutes on an un-named glacier covered 'deep with snow', causing Ruskin to speculate in his Diary as to what he might have said three years previously if he had been told that on that day he 'should climb without fatigue to the aiguille de Charmoz, passing over 'what Saussure[1] himself mentioned with the utmost dread, a glacier covered with fresh snow.' Thereafter, the return was easy, although very wetting as might be expected with the deep snow.'

Editor's botanical note

Despite the distance covered, no specimens were collected on June 13th and only three on the 14th, perhaps because of the snow, two of these being from the high tops and the third from the lower pastures on the walk down.

1. Horace-Bénédict de Saussure, Swiss Alpine explorer and geologist.

Transcript of Ruskin's page 28

ANDROSACE CARNEA. L

I only found this lovely plant once, during all my Alpine rambles, and that was on the 14th of June, in ascending the steep, short-turfed slopes above the Montanvert in order to reach the base of the aiguille Charmoz. It was growing not far above the Montanvert, being when fresh, of a most pure, pale, life like rose colour; and growing in bright delicate clusters of starry flowers – – five leaved.[2] But it lost its colour in drying, totally.

Transcript of verso of Ruskin's page 28

TUSSILAGO ALPINA

This plant I found the same day, June 14th; and I am nearly sure, on the high & barren ridge of rock which runs from the foot of the Petit Charmoz down beside the Glacier de Blaitiere; and on which that day I rested, before my hot struggle up the steep snow of the glacier. It might have been from the lower ridge—but I think not—at all events it was growing in a barren spot, islanded among snows, at 4000 ft above the Prieure.

Editor's note on specimens

'ANDROSACE CARNEA L.'
Now: *Androsace laggeri* A.Huet.
(Pink Rock-jasmine)
Primulaceae (Primrose family)

This quintessentially alpine perennial species grows in meadows and on acid (i.e. non-calcareous) rocks, screes and rocky soils, to an altitude of c. 3000 m. It forms small, loose cushions of leaf rosettes in which the individual leaves are elongated and pointed, almost needle-like, and somewhat hairy, often with bristly margins. The flowers, which appear from June/July to August, have five, 'life like', as Ruskin notes, rose-pink or white petals, and a yellow eye, and are borne on medium- to long-stemmed, flat-topped flower heads (umbels) of up to ten individual blooms.

2. Meaning, with five petals.

COLLECTING DAY SEVEN: RUSKIN'S FLORA PAGE 28

'TUSSILAGO ALPINA'

Now renamed *Homogyne alpina* (L.) Cass. (Purple Colt's-foot), although the specimen is probably of:
Tussilaga farfara L. (Colt's-foot)
Asteraceae (syn. Compositae; Aster or Daisy family)

Homogyne alpina grows on acid soils, often close to snow patches, in wet grassland, among shrubs such as Rhododendron and in open conifer woodland, to an altitude of c. 2900 m. The habitat described by Ruskin, however, is quite different from this. Moreover, the morphological characteristics of *H. alpina* are very different from those of the specimen: it has leaves with a shiny, somewhat leathery upper surface and relatively long, slender flower stems (10–30 cm.), with only one or two leaflets; also, the single composite flower head is constricted at the top of the basal sheath of (sepal-like) bracts to form a goblet shape with the reddish purple florets emerging in a tuft from the neck in June. Only well after Ruskin left Chamonix would the hairy fruits have become visible, as on the specimen.

Tussilago farfara, in contrast, is a pioneer species that prefers wet, open rocky screes, banks and gravelly soils, just as Ruskin describes, to an altitude of c. 2600 m. Also, the specimen has several morphological characteristics that suggest that it is of *Tussilago farfara*. It has three, possibly four, prominent leaflets pressed against the relatively short flower stalk, and a wide-open composite flower head. The florets, which would probably have been golden yellow when formed in March to May, were already developing into hairy fruits when collected by Ruskin in June. Finally, all parts of the specimen, including the leaves and leaf stalks, are hairy, as would be expected of *T. farfara*.

Taken together, these observations suggest that the correct identity of the specimen is therefore *Tussilago farfara*. One anomaly, however, is that the two basal leaves on the specimen are rather small for *T. farfara*. This is probably because they were emerging young leaves that would not have developed into the characteristic large leaves of this species until well after the fruits had been fully set. Moreover, the rather barren terrain in which Ruskin says the specimen was collected might have led to the whole plant, but especially the leaves, being rather stunted.

Transcript of Ruskin's page 29

THALICTRUM AQUILEGIFOLIUM

It is exceedingly common in all the rich pastures on the lower slopes of the hills [and] I preserved it only on account of the elegance of its leaf. It was gathered on the 14th of June, in the steep meadows just opposite the Prieure—on the left of the cascade, as I came down from the glaciers of the aiguille Charmoz.

Editor's note on specimen

'THALICTRUM AQUILEGIFOLIUM'

Thalictrum aquilegiifolium L. (note the slight change to the spelling of the species name; Greater Meadow Rue)
Ranunculaceae (Buttercup family)

This truly elegant and widespread plant grows in wet meadows and woodland margins, often in calcareous areas, to an altitude of c. 3000 m. The specimen, for practical reasons, comprises only the tip of one of the branches of this extremely large perennial plant, which can often be 1 m or more tall. As the name suggests, the compound leaves, each with three notched leaflets, bear a superficial resemblance to the leaves of the common garden plant Columbine (*Aquilegia* spp.). The flowers, which form from May to June on loose, branched inflorescences (panicles) at the tops of the stems, lose their tiny petals almost immediately, leaving only the wonderfully fluffy, delicate pink stamens to delight the eye.

COLLECTING DAY EIGHT: JUNE 17TH, 1844

From the Diary: June 15th–17th, 1844

On Saturday 15th it was too hot to climb, so Ruskin took the charabanc, and after walking through a field or two to show his father the scene of the Pèlerins avalanche, sat drawing near Bossons for three or four hours, impeded only by breezeless heat, and no specimens were collected for the *Flora*.

Sunday 16th was a day of rest and devotion and although no specimens were collected for the *Flora* it was nevertheless botanically interesting. After listening to an excellent sermon, Ruskin sat in his room with 'a branch of sapin'[1] beside him, presumably gathered the previous day 'from one of the topmost fragments of the shattered pines of les Pelerins', which differed 'in many particulars from the lower bough described at Page 5'[2] (see Diary entry for June 12th, p. 93, and accompanying Editor's botanical note, p. 94). Having made drawings of the 'branch of sapin' beside him (p. 104), he then writes as follows:

> In the first place, the main stem, though half an inch thick, retains all the markings of the leaves, as at *a* [drawing apparently missing], the leaves being themselves somewhat stronger and larger than below. The side twigs are the sixth of an inch thick and terminate bluntly without tapering, their leaves being inclined upwards so as to give the section *b* [see illustration overleaf, upper left], which is narrower in the sides than the top owing to the leaves being more curved at the sides; thus being exactly contrary in mode of variation to the lower bough.

In the evening Ruskin retraced his steps of the previous day to the Pèlerins avalanche to study the 'pinetops' once more, and at the same time 'may have caught a little cold'.

On June 17th Ruskin set off at about 8 am with his father and mother to show them the fall of the eau de Bérard, but 'they were neither of them well, lost their tempers and spoiled it all.' The rest of the day was spent more profitably, high above the val

1. Usually translated as 'fir'.
2. The branch in the Diary entry for June 12th is described as 'Mélèze', which is usually translated as 'larch' or '*Larix*', although I identify it as either Silver Fir (*Abies alba* Mill.), or Norway Spruce (*Picea abies* (L.) H. Karst.), tentatively the former.

John Ruskin, drawings of parts of a conifer branch (see Editor's botanical note opposite for details), from manuscript Diary for 1844. Right: This group of drawings, with the upper left corner of the page missing, is opposite the Diary entry for June 16th. Upper left: This single drawing, labelled indistinctly as 'b' *and numbered clearly as* '66', *is on a slip of paper stuck into the front cover of the manuscript Diary; since its indentations appear to fit with those at the top left of the Diary page, it is probably part of the missing corner. This drawing is depicted in Fig. 65 (not 66 as on the drawing) of* Modern Painters V (1860),[3] *so presumably it was cut out to be given to the engraver preparing the printing block. Subsequently it must have been returned to the correct volume, but not the correct page, of the Diary. The whereabouts of the missing drawing* 'a', *which according to Ruskin depicts 'the main stem [which], though half an inch thick, retains all the markings of the leaves …' (see p. 103), are not known.*

d'Entragues, where Couttet pointed out to him 'a flower of the Soldanella which had actually pierced the snow, and was budding in the midst of it.'

They had started at one o'clock to climb the hill that looks into the val d'Entragues, which Ruskin describes as 'a hard piece of work'. They then 'turned aside on the right, to some rocks covered with Alpine rose' (*Rhododendron ferrugineum* or *A. hirsutum*, depending on pH, acidic for the former and alkaline for the latter), from where Ruskin noted the pyramidal form of the Aiguille d'Argentière. Next there followed some steep side work and slopes of snow, the last very steep, which made Couttet nervous for Ruskin's safety; and beyond they got

> a grand view of the Val d'Entragues, filled with snow, and terminated by the

3. *Modern Painters V*, Ch. VIII, 'The Leaf Monuments'; *Works*, vol. VII, p. 96.

> slaty beds and rugged glaciers of the Buet, and cheval blanc. It was a strange scene—one of a kind quite new to me—rendered more impressive by the precipice on which we stood above it, whose edge of alpine rose added to its fearfulness; and by the solemn unbroken misty sunshine of the high air, which came in slanting rays over the vast masses of sloping rounded gneiss stained with a dusty mixture of black and lemon yellow, which formed our foreground. L'Oria rose opposite to us, and the aiguilles Rouges behind. I was amazed at the extent of snow on them and the size of their glaciers. Obliged to descend too soon, though we took a look over a rosy hill into the Val Bérard. This last sterile and lonely, but not impressive. Fancy I have still some signs of cold ...

Despite the area covered, only one specimen, the insectivorous plant *Pinguicula alpina* L. (Alpine Butterwort) was collected for the *Flora*. Interestingly, this species, like the *Soldanella* mentioned at the beginning, prefers alkaline soils, which suggests that for part of the day at least Ruskin was probably traversing calcareous strata.

Editor's botanical note

It appears from the Diary entry of June 16th that the tree branches drawn on June 12th (see entry for that date, p. 93, and for June 16th, p. 103) are from trees at Les Pèlerins. Also, that the branch drawn on the 12th was probably the lower branch of a standing tree and that drawn on the 16th was one of the upper branches of a fallen tree. Neither set of drawings is of a 'pine', as was stated in the Diary entry, but both are of members, or a single member, of the Pine family (Pinaceae). Nor is either set of drawings of a branch of European Larch (*Larix decidua* Mill.), another member of the Pinaceae, as suggested by Ruskin's use of the word *'Mélèze'* on June 12th (see Editor's botanical note p. 94). It is, however, likely that both sets of drawings depict one or other of two further members of the Pinaceae, Silver Fir (*Abies alba* Mill.) or Norway Spruce (*Picea abies* (L.) H.Karst.). It was tentatively concluded in the botanical note for the June 12th Diary entry that the branch depicted there was of Silver Fir.

Ruskin says that the bough from which the June 16th drawings were made 'differs in many particulars from the lower bough described at Page 5' (see entry for June 12th). He also describes the drawings as being based on 'a branch of sapin' (usually translated as 'fir'). And finally, when the drawing from the Diary showing the tip of a branch appears in *Modern Painters* as Fig. 65,[4] it is described as 'the extremity of a single shoot of spruce fir, foreshortened, showing the resilience of its swords [needles, leaves] from beneath ... The fir shoot is carelessly drawn; but it is not worth while to do it better, unless I engraved it on steel, so as to show the fine relations of shade.'

These references suggest that Ruskin identified the tree depicted and described on June 16th as Norway Spruce (*Picea abies*), a conclusion supported by my own examination of the drawings.

4. Ibid.

Transcript of Ruskin's page 30

PINGUICULA ALPINA.

This plant is the most difficult to preserve of any Alpine flowers I have tried. It is full of a fatty liquid, (hence its name),[5] which caused it to take weeks before it dries [sic]—even under the heaviest pressure—and commonly it decays before it can be effectually dried. There is a marked difference, however, between those of the high & low localities. The three uppermost, 1. 2. 3, were from the col above the Val d'Entragues, gathered June 17th, (compare p. 3):[6] there associated with the Soldanella, and growing in the mossy hollows which were wet with the meltings of the snow, at about 2500 ft above the Prieure. These are evidently stronger in stem & firmer in constitution, (keeping their white colour) than the one below, which was gathered June 11th, in the wood; going to the Chapeau, in a very damp spot, (vide p 19.)[7] and which lost its colour, and all appearance of form, long before I could dry it.

Editor's note on specimens

'PINGUICULA ALPINA'

Pinguicula alpina L. (Alpine Butterwort)

Lentibulariaceae (The Butterwort and Bladderwort family)

This fascinating insectivorous[8] plant grows on wet, nutrient-poor, sometimes calcareous soils on the surface of rocks, in rock crevices, on peat bogs and in short grasslands, at altitudes between c. 1500 and 2600 m. The elliptical to oblong/lanceolate, fleshy leaves, with rolled-up margins, form a basal rosette, tightly appressed to the soil surface. The leaves are bright yellow-green in colour, with a moist and sticky surface, usually littered with the remains of small, winged insects. The latter are attracted to the leaf by its colour and glistening surface, and become trapped there by mucilage secreted by glands on the upper surface. The movement as the insects attempt to escape stimulates the secretion of yet more mucilage, together with increasing amounts of enzymes from other glands, to break down their bodies, thereby releasing soluble nutrients that are absorbed immediately to nourish the plant. Only the chitin of the insects' outer skeletons remains as evidence of the dramatic life and death struggles between plant and prey acted out on its deceptively benign leaves.

A clue to the identity of the specimen is provided by Ruskin, who states in his notes that the flowers were white. In addition, although Ruskin fails to mention this, the flowers of this species have two yellow patches in the throat. Other Butterworts, by contrast, have lilac, violet or blue flowers. The form of the Alpine Butterwort flower is asymmetric, the five petals being joined together for part of their length to create a hood, which opens out

5. *Pinguis* is the Latin word for fat.
6. Actually on the lower half of an unmarked page before Ruskin's page 3, which deals with *Soldanella*.
7. This deals with the moss *Polytrichum commune*.
8. i.e. obtains nutrients by trapping and digesting the bodies of insects.

into a three-lobed upper lip and a two-lobed lower lip. At the rear of the flower, the hood comes to a blunt point as a short spur. In one of Ruskin's early attempts to re-classify plants, having rejected the widely accepted approach devised by Linnæus, Butterworts would have been placed in the group which he describes as 'Knight's Hoods Known by the attached Spurs'.[9]

9. Ingram, David [S.] (2016): 'Ruskin's Botanical Books: A Survey of Re-ordered and Annotated Second Edition volumes of *British Phænogamous Botany* (W. Baxter, 1834–43) and *English Botany* (J. E. Smith & J. Sowerby, 1832–40), *Ruskin Review & Bulletin*, 12 (1) (2016), pp. 18–50; and *Ruskin's Botanical Books: Re-ordered and Annotated Editions of* Baxter *and* Sowerby, (Guild of St. George, York, 2016).

COLLECTING DAY NINE: JUNE 20TH, 1844

From the Diary: June 18th–20th, 1844

By June 18th Ruskin's cold had gone and he spent much of the day drawing, getting on 'pretty well, though interrupted by many light showers'. No plants were collected and by the time he wrote up his Diary in the evening, it was 'pouring and the shutters and casements rattle[d] wildly with the wind'.

June 19th was not a good day. The rain had 'plashed ponderously' all the previous night, and although the sun was out until mid-morning, Ruskin almost set off to draw, but luckily 'stopped to set [his] plants in order'. By 11 am it was clear that more bad weather was imminent. He 'ran out to foot of Breven for some plants, Coutet and George following' but was caught by a shower and sheltered, desperately cold, for half an hour, when a gleam of sunshine permitted him to 'run up and get my leaves'. The rain then set in, with brief respites, for the rest of the day. He ended the Diary entry by noting that the mountains were by then 'covered with fresh snow nearly as low as the chalets of the Tapia'.

Despite the beauty of the early morning view on June 20th, it was 'mighty cold' and Ruskin could not draw the pines coming out of the cloud, so he 'sat down to [his] sapins again'. These were the branches he had drawn on the afternoon of Sunday, June 16th (see p. 103), and he compares a fresh specimen with three drawings ('c', 'd' and 'e') which are on the page opposite his Diary entry for June 16th (see illustration p. 104; the drawing 'a' is currently missing from this page, and the drawing faintly labelled 'b' has apparently been cut out and re-fixed at the front of the Diary).

> I brought the top of a dried pine from the forest of the Pelerins; which agrees in most respects with that described at page 18, but it differs in the way the cones grow. On the fresh one they have each a spray to themselves growing at, and from, its extremity, like a handle to a bell rope; but in the dried one—no, I see now—in the dried one they are the spring buds which have exactly the shape of cones, only are much smaller. These spring from the sides of the twigs where a new branch is to be, and appear to be a sort of egg full of nourishment [actually, the tip of a side branch in miniature, before expansion] for the young bough, which appears in the centre of section .c. p. 18. As the bough shoots, the cone seems to expand into filaments [needle-like leaves],

> as is seen on the exterior of .d., whose symmetry however I have lost, for each filament has the regular and perfect form of root .e. [a drawing of an immature leaf], and they are ranged in spiral series [clearly visible in .c. and .d.], leaving black hollows underneath each.

June 20th turned out to be a fine day, so Ruskin 'went and got on badly with [his] drawing at Bossons'. It was 'a lovely spot, a group of blocks, left doubtless by the ancient glacier de Taconay, covered with birches, and all that is beautiful.' A little later, while 'dining' on black bread and honey, and chatting, Couttet told him that 'he sells about 5000 francs of minerals yearly' and then teased George about his fishing. After this interlude they climbed very high up the Brévent in the afternoon, far along through the forest until they came out above the Arve bridge; and as they returned home they were able to enjoy a 'noble sunset'.

Editor's botanical note

The 'leaves', which Ruskin collected with such difficulty on June 19th, do not appear to have been included in the *Flora*. The observations made on the 'sapins' on June 20th, however, are consistent with the identification of the species made in the Editor's botanical note accompanying the Diary entry for June 16th (p. 105), namely Norway Spruce (*Picea abies* (L.) H.Karst.). Finally, on June 20th, five very varied species were collected for the *Flora*.

Transcript of Ruskin's page 31

This plant I found growing here and there sparingly in the woods on the side of the Breven, in my evening walk on the 20th June. It grows very low, one or two blossoms on a stalk. Those below are from the same locality. The wood slopes steeply, to the south east, and the soil is comparatively parched & bare of flowers.

Editor's note on specimens

Upper specimen:
Polygala chamaebuxus L. O. Schwarz (syn. *Polygala chamaebuxus* L.; Shrubby Milkwort)
Polygalaceae (Milkwort family)

This rare, low (5–25 cm high), spreading shrubby species grows in grassland or open woodland at altitudes from c. 900 to 2500 m. Some authors suggest that it prefers calcareous soils. It has alternate, oval to lanceolate, shiny green leathery leaves, usually coming to a sharp point at the tip. The dramatic flowers (c. 1.5 cm long) are borne singly, or in groups of up to three or four, in the leaf axils towards the tips of the branches, and appear from March to July. They have five sepals, the inner ones, the 'wings', being large, petal-like and white, pink or purple in colour. Between them are three true petals and these form a lower 'keel', which is bright yellow.

This species, also collected on the Brévent, is included on p. 25 of Payot's *Herbier des Alpes*.

Lower right-hand specimen:
Scleranthus perennis L. (Knawel)
Caryophyllaceae (Pink family)

Knawel, a tough plant, grows among inhospitable rocks, gravels and sandy, usually acid soils, at altitudes of between c. 700 and 2300 m. It is a low-growing (c. 5–15 cm), tough, spiky-looking perennial with prostrate, creeping branches densely covered with paired, linear, sharply-pointed leaves. The unusual clustered flowers, which form from May to August, lack petals, their role being assumed by five large outer structures (sepals), fused at the base and with membranous edges, which give the flowers a green and white appearance.

Lower left-hand specimen:
Possibly *Cherleria laricifolia* (L.) Lamonico (syn. *Minuartia laricifolia (L.)* Schinz & Thell.; Sandwort)
Caryophyllaceae (Pink family)

From the structure of the flower and the paired arrangement of the leaves, this lonely specimen appears to be a member of the Pink family (Caryophyllaceae) and is probably a single stem from the cushion of a Sandwort. The specimen has a much larger flower, with substantial sepals and petals, and is altogether tougher looking than the delicate specimens of another Sandwort on Ruskin's page 12 of the *Flora*. Moreover, according to Ruskin's collecting notes, it was growing in very poor, dry conditions, whereas the earlier specimens were from a moister area.

The present single stemmed specimen, with a single flower and an opening bud below it, resembles *Minuartia laricifolia*.[1] This perennial, loose cushion-forming species generally grows in rocky, siliceous grasslands at altitudes from c. 1200–2500+ m. The tough flowering stems have paired, narrow and pointed leaves, and one or more large white flowers per stem (non-flowering stems usually also have needle-like leaves in groups of one to six at their tips). The plant is usually described as being hairy, although Ruskin's specimen appears to lack hairs. However, the presence of hairs may be a variable feature, since some specimens of *Minuartia laricifolia* in the Herbarium of the Royal Botanic Garden Edinburgh are glabrous (smooth) and others are only slightly hairy.[2]

1. I thank Dr Alan Elliott of the Royal Botanic Garden Edinburgh for expert advice concerning the possible identity of Ruskin's specimen.

2. Personal communication: Dr Alan Elliott, Royal Botanic Garden Edinburgh.

Transcript of Ruskin's page 32

VERONICA URTICIFOLIA.

The largest and darkest had lost its number when I unpacked it; but is I believe, (the other is certainly) from the same woods as the preceding. A different group of flowers appears to me to grow on this part of the Breven, from those of any other locality in the valley.

Editor's note on specimens

'VERONICA URTICIFOLIA.'
Veronica urticifolia Jacq. (Nettle-leaved Veronica)
Plantaginaceae (Plantain and Veronica family)

This dramatic, large (20–70 cm tall), slightly hairy Veronica grows in moist, coniferous woods and other shady places, usually on calcareous soils, from an altitude of c. 900 to 2300 m. The paired leaves, which are somewhat nettle-like, lack stalks (petioles) and are oval to lanceolate, almost heart-shaped in outline, with sharply toothed margins, the teeth pointing towards the tip. The flowers, which appear from June to August, are borne in long spikes, which grow from the axils of opposing leaves. Each flower has four pale lilac to pink petals, joined together, and two prominent stamens.

Transcript of Ruskin's page 33

GNAPHALIUM DIOICUM.

These specimens, which were the finest I ever found, are from the open heights which I reached the same evening, after passing through the wood; and about half way up the Breven.

Editor's note on specimens

'GNAPHALIUM DIOICUM.'
Now *Antennaria dioica* (L.) Gaertn.
(Mountain Everlasting or Cat's-foot; female plants)
Asteraceae (syn. Compositae; Aster or Daisy family)

Antennaria dioica is 'dioecious' (having male and female flowers on different plants), hence the specific name. Specimens of male plants may be seen associated with Ruskin's p. 3 of the *Flora*, with a detailed description of the species as a whole in the accompanying Editor's botanical note (p. 60).

The flower heads of female plants, which Ruskin has collected here, are usually relatively large and far more spectacular than those of male plants. They are borne on tallish stems and not only have pink disc florets in the centre, but are also surrounded by pink bracts. Close examination of the specimens reveals traces of this pink colouration still remaining. It is no wonder that Ruskin described these striking plants as 'the finest [specimens of this species] I ever found'.

COLLECTING DAY TEN: JUNE 21ST, 1844

From the Diary: June 21st, 1844

June 21st was 'Cloudless but not too warm, which was well, for we had a tremendous climb; nothing but *mauvais pas* from beginning to end, and a little too steep to be pleasant.' Moreover, Ruskin was disappointed at the top, 'for though the view of the valley was very noble, we seemed no nearer the great precipice on the Dome du Goute than we are here; and I felt myself in every way deceived by the size of objects. These mountains are just like St. Peter's, proportioned to littleness.' In the Diary Ruskin continued to reflect on the optical illusions experienced in the mountains by noting that he 'was struck also by a deception new to [me], that steep slopes, seen from a great height above them, look almost level'. He unwisely took a little wine at the top, which gave him a dreadful headache in descending and at the bottom he felt quite ill. However, although frightened, he walked as fast as he could through the hamlet of Bossons, which apparently cured him.

It was a successful day, botanically, with six very varied species being collected for the *Flora*, all probably found on acidic to neutral substrates.

Transcript of Ruskin's page 34

Half way up the Montagne de la Cote; June 21. we sat down to rest on the first bit of ground we had come to, that afforded more than a step for the foot, and a ledge for the hand. It was at the top of a small cascade, which slid for some yards trickling and glittering, over a rounded slope of smooth rock, and then fell from the edge of it in divided streamlets, nearly down to the Glacier de Taconay [sic]. In the middle of the stream, and in the crannies of the rock, grew this plant, associated with the flowers of the R. glacialis (vide p 45) preserved with it, and which were the first I had ever seen.

Editor's note on specimens

Centre specimen:
Neogaya simplex (L.) Meisn. (Syn. *Ligusticum mutellinoides* (Crantz) Vill.; Unbranched Lovage)
Apiaceae (syn. Umbelliferae; Carrot and Cow-parsley family)

This rather stunted, stocky plant normally grows on acid screes and rocky or stony grasslands to an altitude of c. 3000 m. The plants are almost hairless, 10–25 cm tall, un-branched and have only basal leaves. These are soft, grey-green in colour, aromatic and compound (two to three pinnate), with the leaflets having long, narrow lobes. The flowers, which may be creamy white through to deep pink in colour, appear from June to August, and are borne in flattish heads (umbels) with large, three-toothed basal leafy structures (bracts) that are retained until after fruiting.

This species, identified as *Gaya simplex* Gau., is included on p. 19 of Payot's *Herbier des Alpes.*

Four corner specimens:
Now *Beckwithia glacialis* (L.) Á. Löve & D. Löve (formerly *Ranunculus glacialis* L.; Glacier Buttercup or Glacier Crowfoot)
Ranunculaceae (Buttercup and Crowfoot family)

These are incomplete specimens. For full details, see Editor's botanical note (p. 131) on the splendid, complete specimen on Ruskin's page 45 of the *Flora of Chamouni 1844*.

Transcript of Ruskin's page 35

SAXIFRAGA. BRYOIDES

This delicate little flower I have marked only as from the side of La Cote, June 21, nor do I recollect its exact locality.

Editor's note on specimens

Upper specimens:
'SAXIFRAGA. BRYOIDES'
Saxifraga bryoides L. (Mossy Saxifrage)
Saxifragaceae (Saxifrage family)

This diminutive Saxifrage usually grows in wet, lime-poor siliceous soils on stony ground and among rock and screes, sometimes near snow patches, at altitudes of between c. 1800 and 3000+ m. The plants form mats comprised of moss-like rosettes of narrow, incurving leaves edged with bristly hairs. A single long, slightly hairy flower stem bearing a large star-like flower arises from each rosette from June/July to August. Each flower has five short sepals alternating with long, white or creamy-white petals, yellow spotted or blotched towards the centre, with distinct, radiating stamens.

Lower specimens:
Hornungia alpina (L.) O. Appel (Chamois Cress)
or, depending on the pH of the substrate (see below)
H. alpina subsp. *brevicaulis* (Spreng.) O. Appel
Brassicaceae (sometimes referred to as Cruciferae; Cabbage family)

Chamois cress, a small (c. 3–15 cm) perennial plant, grows in the harsh environment of wet limestone screes and rock crevices and hollows, at altitudes of between c. 2000 to 3400 m. However, a variant[1] (*H. alpina* subsp. *brevicaulis* (Spreng.) O. Appel), that has slightly smaller flowers than *H. alpina*, is said to occur in similar habitats on basic or acidic siliceous rocks.

The somewhat fleshy, hairless leaves of both the species and subspecies form a basal rosette and may be either small and undivided, or divided into three or often more deep lobes in a pinnate (feather-like) form. The small flowers, borne on an upright, largely leafless, hairy stem (hairless in ssp. *brevicaulis*), branched to form a corymb, have four white petals and appear from May to August. The seed pods (siliquae) are dry, elongate-oval and flattened in form, pointed at the tip, as in the upper of the three specimens (blunter-ended in ssp. *brevicaulis* according to some authors).

1. Grey-Wilson & Blamey (1995), p. 90; Ménard (2007), p. 102.

Transcript of Ruskin's page 36

Also from the side of La Cote, June 21. I was so fatigued when I came down that evening that I fell asleep in writing; and never noted the exact localities.

Editor's note on specimens

Upper specimens:
Possibly *Luzula lutea* (All.) DC. (Yellow Woodrush)
Juncaceae (Rush family)

This relatively small, erect Woodrush, common throughout the French Alps, prefers dry acidic grasslands and moors at altitudes between c. 1700 and 3000 m, where it spreads by means of underground stems (stolons). Unlike other Woodrushes, it is not at all hairy, and the grass-like leaves are linear and soft. The tight, pale golden yellow flower clusters (glomerules), each borne on a long, flexible stalk, are formed in loose panicles at the tops of the upright stems. The flowering period is said to be from July to August, so it would seem that Ruskin's specimens were flowering a little early.

Lower specimen:
Erigeron uniflorus L. (One-flowered Fleabane)
Asteraceae (syn. Compositae; Aster or Daisy family)

This small (c. 2–10 cm), but robust-looking tufted perennial species grows in grasslands on exposed ridges, in rock crevices and among rocky debris, often near snow patches, at altitudes of c. 1200 to 3500 m. The clustered, hairy leaves have a spatula-like form, broadest towards the tip. The stem is upright and smooth at the base, becoming hairy towards the solitary, composite flower heads that form from June/July to August. Each flower head comprises a centre of yellow disc florets surrounded by pinkish-lilac to white ray florets, the whole structure being enclosed by upright leafy structures (bracts) which, being woolly, appear mainly white, but with lilac tips.

This species is included on p. 36 of Payot's *Herbier des Alpes*.

COLLECTING DAY ELEVEN: JUNE 22ND, 1844

From the Diary: June 22nd, 1844

'By far the most delicious day I have had yet', wrote Ruskin, setting out after a late start for the foot of the Aiguille du Midi. After passing the avalanche, a winding path led them up beside one of the gorges of the glacier stream … 'which is a noble gorge, bordered by precipices and at present deeply filled with massy snow. Rich copsewood about us; alpine rose; showing through its gaps a steep slope of meadow deep in grass and dew, the slanting morning light giving it all value.' They then had a winding climb among turf and alpine roses for some fifteen hundred feet, which brought them to a ridge overlooking the Glacier des Bossons. After resting, they had a steep sideways scramble to the edge of glacier, which Ruskin declared to be 'the most impressive bit of ice, from its vast size and height, I have yet seen. In the hollow of its moraine, from a block recently shattered by the fall of another upon it, I got the bit of granite marked'. Then a stiff ascent of a steep gulley, enlivened by Couttet offering advice on the best way of avoiding any stones which might fall from above, led to a noble view of the valley on both sides.

After dining under a huge stone, looking alternately to the Brévent, the Aiguille du Plan and Mont Blanc, another half hour of stiff climbing brought them to a chasm where Couttet went down and collected a specimen of *Saxifraga oppositifolia* (see Ruskin's page 39, this edition of the *Flora*). They then had a difficult 30 minutes on a moraine to the edge of the Glacier des Pèlerins, at a point which Ruskin thought was higher than any he had previously reached, from where he could see 'the Chapeau and the rise above Les Lines over the top of the Tapia'. After spending 'a solitary half hour here of great value' to him, he descended the moraine, and found a block containing a large vein of quartz, with what seemed to be one of granite in the middle of it. Incautiously, he tried to roll it down the hill, but it took a wrong turn and was lost. After arriving back at 'our dining stone' he lay down in the shade of it and slept for half an hour, before descending as fast as he could. This nevertheless took two hours, with Ruskin boasting at the end that he was 'as fresh as if I had done nothing'.

This strenuous and exciting day yielded six varied and interesting species for the *Flora*, some preferring alkaline and others acidic soils or substrates. One of the species gathered from the very highest point reached was a 'noble specimen' of *Ranunculus glacialis* L. (Glacier Crowfoot; now *Beckwithia glacialis* (L.) Á. Löve & D. Löve). This was not, however, mounted in the *Flora* with the other specimens gathered on June 22nd, but was 'reserved ... for the last [page of the *Flora*, Ruskin's page 45], because it is the most thoroughly characteristic of the Alps—growing in the most desolate places, with nothing but the moraine sands, wet with ice streams, for a soil. It haunts the borders of the glaciers in a zone of about from 5 to 9000 ft above the sea, seldom occurring lower, nor far from ice ...'

Transcript of Ruskin's page 37

On the 22nd of June, in climbing to the foot of the aiguille du Midi, I found this plant, either in the thicket at the bottom, near the glacier des Bossons, which I rather think, or else in the open meadows just above.

Editor's note on specimen

Epilobium fleischeri Hochst. (Alpine Willowherb)
Onagraceae (Willowherb family)

This compact and flamboyant perennial, which overwinters as buds just below soil level, grows in acid moraines, effluvium from glacial outfalls and gravelly soils at altitudes of between c. 500 and 2500 m. The alternately arranged, densely packed, slightly notched and hairless leaves are borne on stems varying in height from 10 to 30 cm. The striking bright pink or red flowers, which form at the tips of the stems from June/July to August, have four thin, pointed, purplish sepals, four rounded petals without notches, and prominent stamens. The long, narrow fruit pods split when ripe to release the seeds which, having long silky plumes, are easily borne far and wide on the wind.

Transcript of Ruskin's page 38

LINARIA. ALPINA.

This lovely plant, though I do not imagine it is uncommon, I have yet only found twice among The Alps. The first time, this day, as we ascended from the border of the glacier des Bossons, through the couloir, to the shoulder of the aiguille.

The watercourse was the steepest bit of continuous ascent I climbed in Chamonix[.] For twenty minutes we could not loose hold with our hands without danger, and there was some chance also of stones coming from above, so that we hardly stopped to breathe. The soil too was dry & dusty, and loose for the foot, & the only redeeming point of the whole was, that this flower, then new to me,—was growing in large bunches of the most intense purple—all about the crannies of the loose soil. The second time that I found it was on the very highest turn of the Simplon road, the evening that I walked up from the village to see the sunset on the Aletsch horn,[1] ( )[2] there also growing on a dry, dusty soil to which its roots hardly adhered. There I found two species, one like this, with the vivid orange centre, the other having this orange replaced by a pale lilac. In both these situations the plant was growing at more than 6000 ft above the sea.

Editor's note on specimen

'LINARIA ALPINA'

Linaria alpina (L.) Mill. (Alpine Toadflax)

Plantaginaceae (Plantain family; previously included in the Scrophulariaceae [Figwort family])

Alpine Toadflax,[3] a delightful annual or biennial[4] high altitude pioneer species, prefers damp limestone moraines and bare rocky, stony or gravelly soils or grasslands, at altitudes of c. 1500–3500 m. The small plants are low-growing or prostrate, with many elongate-oval, slightly fleshy, greyish-green leaves in whorls of three or four. The flowering stems are erect and each holds aloft one or two relatively large and striking, viper-like, intense violet-purple flowers. These have a long, curved spur at the rear, and from the front appear open-mouthed and two lipped, the lower lip forming a blister-like, vivid orange 'lump in the throat'. In some genetic variants, as Ruskin notes, this lump lacks the orange pigment and is instead a no less beautiful, delicate pale lilac colour.

1. Aletschorn, the mountain in Switzerland.
2. Brackets left empty by Ruskin.
3. It is interesting that a few years later Ruskin was to be attracted by the related *Cymbalaria muralis* (the prosaically named Ivy-leaved Toadflax, or more romantically, Erba della Madonna), which he found scrambling over the old stonework of Venice.
4. Annual: growing anew from seed every year. Biennial: growing anew from seed in the first year and then, after overwintering in a vegetative state, flowering, seeding and finally dying in the second year.

Transcript of Ruskin's page 39

This I believe to be an exceedingly rare plant. I never saw it but once, and then I could not get at it myself, but Coutet scrambled down into the chasm & brought it up for me,[5] indeed he was the first to point it out. It was in a rocky cleft, just at the base of the aiguille du Midi; near the point of entering on the glacier in ascending Mont Blanc; about 500 ft above the couloir in which I gathered the preceding plant—Its colour when fresh, is an intense smalt—a mixture of lake and ultramarine; exquisitely beautiful.

Editor's note on specimens

Saxifraga oppositifolia L.[6] (Purple Saxifrage)
Saxifragaceae (Saxifrage family)

This tough and adventurous little plant with very large flowers grows on damp, often calcareous[7] rocks, rock crevices, screes and stony soils or grassland at altitudes between c. 800 and more than 3000 m. Hoppe[8] suggests that it holds the European altitude record of 4505 m and *Wikipedia*[9] claims that it grows on Kaffeklubben Island, Greenland which, at 83° 40' N, is the most northerly flowering plant location in the world. It grows as loose cushions, c. 4–10 cm high, with a trailing habit. The small (c. 2–6 mm), elongate-oval leaves, bordered with stiff eyelash-like hairs, are grey-green in colour, often with chalky encrustations, and are arranged in four rows on the thin, woody stems. The relatively large flowers (c. 1–2 cm across), which form in June or July, are solitary and borne on very short stems. The colour of the five petals, which are about three times as long as the sepals, is variable and complex and has been described by different authors as red or violet, purple or lilac, and pale pink to deep purple, while the prominent anthers have a bluish tinge. Ruskin's description of the flower colour as 'when fresh, … an intense smalt—a mixture of [crimson] lake and ultramarine; exquisitely beautiful', is perfect. The species is, however, much more common than his phrase 'exceedingly rare' might imply. This is a plant that grows in extreme environments, so perhaps it was simply 'rare' in the less extreme places where Ruskin had been collecting previously.

5. See the Diary entry for June 22nd (above, p. 118).
6. I thank Dr Alan Elliott, Royal Botanic Garden Edinburgh, for his advice concerning the identity of this specimen.
7. The only rock types mentioned by Ruskin in his Diary entry for June 22nd, granite and quartz, are acid and there is no evidence that the rocks at the bottom of the chasm where this specimen was collected were calcareous. However, although preferring calcareous rocks and soils, this species is definitely not restricted to them (Jim Jermyn, personal communication).
8. See list of field guides in 'Alpine Botany and the *Flora of Chamouni 1844*' pp. 38–9.
9. When accessed during August 2018.

Transcript of Ruskin's page 40

This plant is certainly one of the hardiest of the Alps. I found it growing in large bunches, June 22nd, among the loose blocks of the uppermost moraine of the glacier des Pelerins, at a height of nearly 9000 ft above the sea, close to the ice, and associated only with the Ranunculus Glacialis. Comp. p. 45.

Editor's note on specimens

Upper specimens:
Androsace alpina (L.) Lam. (Alpine Rock-jasmine)
Primulaceae (Primrose family)

This rare, delightful and very tough, low-growing perennial forms turf-like mats on acidic siliceous and granitic rocks and moraines near the high summits of the Alps, at altitudes between c. 2000 and 3000+ m. The species is invariably found associated with *Ranunculus glacialis*[10] (now *Beckwithia glacialis*[11]), as Ruskin describes. The stems are short, tightly packed together and clothed with small (2–10 mm), oblong to lanceolate, grey-green leaves covered with short, starry hairs. The small (7–9 mm) flowers, which appear from June to August, are solitary, held on short stems above the leafy rosettes and have five pointed green sepals and five rounded pink or white petals, with a yellow eye at the centre of the flower.

Lower specimens:
Possibly *Saxifraga moschata* Wulfen (syn. *S. muscoides* Wulfen; Musky Saxifrage)
Or alternatively:
Saxifraga exarata Vill. (White Musky Saxifrage)
Saxifragaceae (Saxifrage family)

The species of *Saxifraga moschata* Wulfen and closely related *Saxifraga* spp. and sspp. are very variable with regard to the number and form of the leaf lobes, the presence or absence of leaf furrows or leaf hairs, flower colour (pale yellow-green or white), and the pH of the preferred substrate.

10. Jim Jermyn, personal communication.

11. See Editor's note to Ruskin's page 45 of the *Flora*, p. 131 below.

COLLECTING DAY TWELVE:
JUNE 23RD (SUNDAY)–JUNE 24TH, 1844

From the Diary: June 23rd (Sunday)–June 24th, 1844

The botany-free Sunday, June 23rd, was spent quietly by Ruskin, with reflective contemplation of the beauty of the scene about him, a 'short saunter' with his father 'in the forenoon up the road to the Montanvert and back by the rocky bases of hills', and in the afternoon a 'meditative walk' to the source of the Arveyron.

The following day, June 24th, was 'Desperately hot. ... the atmosphere like that of an oven.' Ruskin therefore spent the morning near the source of the Arveyron, chiefly in drawing saxifrage leaves among the conifers, for the light from the stones was too strong for his eyes. After 'dinner' he had a tiring scramble up the Montenvers, above the forest of the Arveyron, where he saw *Oxalis acetosella* L. (Wood Sorrel) in great profusion and perfection among the rocks. But it was not a good day, he recalled later, his 'memory seeming to get worse and worse'. Finally, a climb up by the Glacier de Taconnaz was lonely and solemn, and although parts of the glacier itself were interesting, scarcely a moment passed without some stone losing its balance on the moraine and bounding down. 'George' Hobbs let his coat and pole slide down to the glacier, and the safety-conscious Couttet was anxious the whole time they were up, apparently dreading the descent, but eventually they all managed to get down safely.

Despite the heat and other challenges, the day yielded three interesting and contrasting plant specimens, which were ultimately mounted together on a single page of the *Flora*.

Transcript of Ruskin's page 41

ASTRANTIA MINOR.

This plant is a very characteristic feature of the lower rock foregrounds of Chamouni. This specimen, a sufficiently bad one, was gathered June 24th, by the road side in the ravine by which the Arve leaves the valley. Its leaf occurs in a foreground study made on the Montanvert. The Veronica, from the same place, is the most frequent species of Chamouni. It occurs in greatest perfection about the Tete Noire.

Transcript of verso of Ruskin's page 41

This plant was gathered June 24th, in the flat marsh which forms the foreground of my sketch with the pines, near les Ouches [sic]. I preserved it because Coutet affirmed it to be the best remedy known, against fevers.

Editor's note on specimens

Upper right specimen:
Astrantia minor L. (Lesser Masterwort)
Apiaceae (syn. Umbelliferae ; Cow Parsley and Celery family)

This most delicate perennial plant may usually be found growing on acidic, nutrient-poor soils in grassland, open woods and rocky places at altitudes of between c. 1000 to 2700 m. The leaves, which are mostly basal, are deeply divided, almost to the centre, into 5–9 lanceolate lobes with toothed margins. The branched flower stems, which appear from July to August, are from 20–60 cm high with each branch bearing at its tip a compressed, branched and flat-topped flower head (an umbel), 1–1.5 cm diameter. The individual flowers are minute, white to pink in colour and clustered in the centre of the head, which is surrounded by large, pointed papery bracts that overall have a greenish appearance, but are whitish on the underside and often flushed very slightly pink above.

Upper left specimen:
Veronica fruticans Jacq. (Rock Speedwell)
Plantaginaceae (Plantain family; formerly a member of the Scrophulariaceae [Figwort family])

This low growing but striking perennial is usually found on dry, stony soils, loose grassland or other rocky places to altitudes of between c. 1000 and 3000 m. The branched stems are smooth and slightly woody at the base, becoming finely hairy towards the apex. They bear pairs of thick, somewhat oval and shiny leaves, which are only slightly toothed or may even lack teeth and have very short petioles. The small but dramatically coloured, wide-open flowers form in terminal clusters from June to August. Each has a deeply divided, five lobed outer ring of sepals, four intensely blue petals, fused at the base, with a red ring surrounding the throat, and two prominent, protruding stamens.

Lower specimen:
Menyanthes trifoliata L. (Bogbean)
Menyanthaceae (The Bogbean family)

Although no flowers are included, the trifoliate leaves collected by Ruskin, which bear a superficial resemblance to the leaves of the unrelated Broad Bean, are characteristic of *Menyanthes trifoliata*, a conclusion confirmed by the boggy location. In fact, Bogbean may occupy a diversity of wet sites including bogs, mires, marshes, pond margins and ditches, to an altitude of c. 1800 m. It has stout, under-water creeping stems (rhizomes) that produce smooth, fleshy leaves and lax flower spikes (racemes), which are held well above the surface of the water to a height of c. 15–30 cm. The distinctive, star-like flowers, which form from May to June, have five green sepals and five pointed petals that curve backwards. These are white below and a delicate soft shell-pink above, and are fringed on the upper surface with long, white hairs. The leaves and rhizomes, although poisonous, have been and still are used as traditional medicines and food and drink additives around the world.[1]

1. See, for example, Mabberley (2017), p. 579.

COLLECTING DAY THIRTEEN: JUNE 25TH, 1844

From the Diary: June 25th–27th (Collecting day 13)—August 3rd, 1844

On June 25th it rained in the morning, but half cleared at 10 am, so Ruskin took a charabanc and 'botanized at the gorge of the Arve on the Servoz road, sketched a little, and was driven home by rain.' Whether any plants were collected is not known, but none survived to be included in the *Flora*.

June 26th started well at 4.15 am, with Ruskin declaring 'Of all the lovely dawns I ever saw on Mont Blanc, this bears the bell'. However, the morning deteriorated rapidly into 'a disagreeable windy bleak day for drawing: worked hard and spoiled all as usual.' In a short ramble to warm himself he found 'a glorious little bit of scenery by the Arve side, a meadow under gneiss[1] cliffs, with a broken cascade gushing into it and dark pines separating it from the roar of the river.' Moreover, he had 'a delicious climb too with Couttet, among these same gneiss rocks, which bear most clear and indisputable evidence of the former passage of glaciers over them.' Again no plants appear to have been collected.

June 27th was also 'Rather a failure of a day', with a broken, ugly grey sky in the morning when Ruskin went with his father and mother to show them the dingle at the foot of the Midi, intending to draw afterwards at the avalanche, 'but it came on pouring rain, and I had nothing for it but ordering my plants in the afternoon.' 'I believe I shall have some pleasure out of them' he wrote in his Diary that evening, as indeed he did if the *Flora of Chamouni* is anything to go by. Despite the rain, three plant species, albeit from the valley rather than the mountains, were collected for the *Flora* during the day.

Sadly, there was to be no more plant collecting during the 1844 visit to Chamonix. The weather remained unsettled, and the rest of the time appears to have been taken up with short outings between spells of rain, sometimes to draw, and family cares, Ruskin's father and mother both being unwell at times. On July 3rd he opened his Diary and wrote: 'CHAMOUNI. I write this loved name for the last time at present, and am very low tonight. I love the valley better and better every day and have not done the thousandth part of what I hoped. The weather has been against me this last fortnight.'

After visiting Baveno on Lake Maggiore, and then Zermatt, the Ruskins were in fact back in Chamonix by the 25th, but this visit was a very short one and on August 3rd they began their journey to Paris, and eventually home to England.

1. A coarse, foliated, metamorphic rock.

Transcript of Ruskin's page 42

These specimens were gathered June 27th, close beside the bridge which forms the lowest part of the pass of the Tete Noire; and from which the road ascends to the gallery.

Editor's note on specimens

Epilobium montanum L. (Mountain Willowherb)
Onagraceae (Willowherb family)

Mountain Willowherb is a medium height perennial (c. 25–80 cm) and is usually found on loose soils, often shaded, at altitudes from 1000 to 2500+ m. The stems, normally branched near the base, are smooth or very finely hairy. The ovate to lanceolate leaves are coarsely toothed and are carried in opposite pairs lower down the plant, becoming alternate higher up. The bell-like flowers, carried at intervals up the stems, appear from May to August, each having four purple-pink petals and a four-lobed stigma. They form at the apex of elongated ovaries, which mature after fertilization into long, slender, fruits (capsules) which, when dry, split lengthwise to release numerous fluffy seeds that are readily dispersed by the wind.

The specimens are not complete and the flowers closed, so the species cannot be identified with certainty. It is possible, therefore, that they may be of *Epilobium collinum* C.C.Gmel. (Hill Willowherb), a species similar to *E. montanum*, but somewhat shorter, with slightly paler and later opening flowers.

Transcript of Ruskin's page 43

THYMUS ALPINUS

From the same spot as the preceding species.

Editor's note on specimen

'THYMUS ALPINUS'
Thymus alpinus has now been renamed as:
Clinopodium alpinum (L.) Kuntze
(Alpine Calamint or Rock Thyme)
However, it is often referred to in books using the following synonym:
Acinos alpinus (L.) Moench (Calament des Alpes)
Labiatae (syn. Lamiaceae; Deadnettle family)

Alpine Calamint is a relatively small (10–30 cm), tufted, somewhat hairy perennial that occurs in rocky or stony grassy areas and screes, often but not exclusively on limestone, at altitudes between c. 1200 and 2700 m. The paired leaves are oval to slightly pointed, usually sparsely toothed, especially towards the apex, with a short stalk (petiole). The reddish-purple flowers (1–2 cm long), which form in small leafy whorls near the tips of the stems from late June until August, are tubular, widening to form a dragon-like mouth with pronounced upper and lower lobed lips, the inside lower lip bearing semi-circular white markings. Some authors describe the plant as odourless, others as having a minty scent when crushed.

The species is included in Payot's *Herbier Des Alpes*, p. 28, labelled as '*Thymus alpinus*, Brevent'.

Alpine Calamint is sometimes confused with the related, but annual, Alpine Basil-thyme or Calament des Champs (*Clinopodium acinos* [L.] Kuntze; syn. *Acinos arvensis* (Lam) Dandy).

Transcript of Ruskin's page 44

Gathered on the rocks, by the side of the path, a little nearer Chamonix than the preceding plants. The rocks were the "poudingues de Valorsine" of Saussure,[2] and the situation close to the arch which marks the boundary of Savoy & Switzerland.

Editor's note on specimen

Medicago lupulina L. (Black Medick)
Fabaceae (syn. Leguminosae; Bean or Clover family)

This elegant annual occurs widely on short grasslands, often but not exclusively over limestone, and on well-drained soils or verges, at altitudes up to c. 2400 m. The plants are low, sometimes prostrate to sprawling, usually downy, with alternate, trifoliate leaves on long petioles with pointed leaf-like structures (stipules) at the base. The leaflets are oval or rounded, wide towards the apex, sometimes almost heart-shaped, and are usually finely toothed with a small apical point. The tight, convex to flattish flower heads (racemes), comprised of many small (2–3 mm), clear yellow flowers, form on long, slender stalks in the angles of leaf stalks and stem (axils) and appear from April until September. As the kidney-shaped (almost coiled) pods ripen to black, the flower heads elongate to accommodate them.

2. Ruskin is probably referring to a kind of schist or gneiss (medium grade and high grade, respectively, foliated metamorphic rocks that split into sheets to reveal pieces of such minerals as mica, feldspar and quartz) that occurs in the Vallée de Valorsine and contains extraneous pebbles and larger stones up to the size of a head, which the Swiss geologist Horace-Bénédict de Saussure (1740–1799) referred to as 'poudingues' (pudding rocks). See H. B. de Saussure, *Voyages dans les Alpes*, Vol. 1, (Samuel Fauche, Neuchâtel, 1779); and Wilson, J., *A History of Mountains Geographical and Mineralogical*, Vol. 2, (Nicol, White, Faulders, and Asperne, London, 1809), pp. 407–408. Ruskin had in his library both the 1779–96 and the 1803 editions of Saussure's work: see Dearden, J. S. (2012), Cat. nos. 2373 & 2374.

Transcript of Ruskin's page 45

RANUNCULUS GLACIALIS

I have reserved this flower for the last, because it is the most thoroughly characteristic of the Alps—growing in the most desolate places, with nothing but the moraine sands, wet with ice streams, for a soil. It haunts the borders of the glaciers in a zone of about from 5 to 9000 ft above the sea, seldom occurring lower, nor far from ice (compare however p. 34.) This which is a noble specimen, (the plant usually growing low & in bunches of blossoms with a stalk to each, and seldom in this branch like form), was gathered June 22nd, from the sandy slopes of the moraine of the Gl. des Pelerins (comp. page 40) where it was growing in great profusion. It is said to be very medicinal.[3] Coutet always gathered & carried down with him all he could find.

Editor's note on specimen

'RANUNCULUS GLACIALIS'
Beckwithia[4] *glacialis* (L.) A. Löve & D. Löve
formerly *Ranunculus glacialis* L. (Glacier Crowfoot)
Ranunculaceae (Buttercup and Crowfoot family)

This most beautiful and fitting plant to end the *Flora* inhabits some of the highest peaks in the Alps. It may be encountered in wet, inhospitable, acidic rocky habitats, including glacial moraines and screes, often close to late snow patches, at altitudes from c. 1700 to over 4000 m. It is perennial, relatively small and stocky, with only a single stem or just a few branches, and reaches a height of anything between c. 5 and 20 cm, depending on altitude and aspect. The leaves are large, smooth and fleshy, and mostly deeply divided, especially those at the base, with an overall palmate form. The magnificent flowers, which appear from June/July until August, may be borne singly (see, for example, the specimens next to Ruskin's page 34) or in small groups, as here, and are large (c. 3–4 cm diameter) and long lasting. The bud scales (sepals) have reddish-brown hairs, but the five petals, although often pure glistening white, are also frequently pink or even a dramatic red-purple colour. It is usually assumed that the colour of the petals changes from white to pink and reddish shades following fertilization or as a result of ageing, but I suggest that it may equally well result from natural genetic variation within the species.[5]

3. For example, it was said to be 'sudorific' (induces sweating, thereby alleviating the symptoms of disease) by A. D. Turnbull, M.D.: *On the Medical Properties of the Natural Order Ranunculaceae*, (Longman *et al*, London, 1835) p. 3.
4. After Edward P. Beckwith (1887–1966), a North American engineer, chemist, climber (especially in the Alps and Dolomites) and explorer, who died at the age of 89 in 1966 (*The New York Times*, July 7th, 1966).
5. Jim Jermyn has dissected unopened buds of this species and found the red pigmentation of the petals already present (personal communication). He also reports that he has seen inter-generic crosses of *Ranunculus glacialis* [sic] and *R. parnassifolius* L. with red-pigmented petals present at the time of flower opening.

RUSKIN AND BOTANY

The *Flora of Chamouni* probably represents the end of John Ruskin's botanical age of innocence. In the decade following 1844, he gradually became more and more disillusioned with the work of conventional botanists, especially taxonomists – those who classify and name plants – and in 1855 began the process of formulating increasingly idiosyncratic schemes of his own. This culminated, as demonstrated by the following chronology, in the writing of *Proserpina*, his remarkable book on botany described by Wilfrid Blunt and William Stearn[1] as 'the most stimulating book ever written about flowers', but by Richard Mabey as 'a confused and at times deranged attempt to devise a new, anti-Linnæan plant taxonomy, based on æsthetic principles rather than scientific understanding'.[2]

A Short Chronology of Ruskin's Botanical Studies

CHILDHOOD The young John Ruskin's interest in plants and flowers began when, after finishing his morning task of learning passages of the Bible or Latin grammar, he would escape to the garden of the family home at Herne Hill, where his mother, Margaret Ruskin, was 'often planting or pruning beside me'[3]. Later, he wrote: 'Constantly … in the garden … my time there was passed chiefly in the … close watching of the ways of plants … staring at them, or into them. In … admiring wonder, I pulled every flower to pieces till I knew all that could be seen of it with a child's eyes …'[4].

After this the scene shifts to other memories, set out in the introduction to Ruskin's book on botany, *Proserpina*,[5] where he wrote: 'Yesterday evening I was looking over the first book in which I studied Botany,—Curtis's [Botanical] Magazine, published in 1795'; then, after a long account of the many deficiencies of botanical nomenclature and 'Linnæan' classification he goes on to state: 'but—although I know my good father and mother did their best they could for me in buying this beautiful book; and though the admirable plates … taught me much, I cannot wonder that neither my infantine nor boyish mind was irresistibly attracted by the text'.[6] Thus the seeds of both serious botanical study and criticism of conventional approaches to classification, which were to develop into controversial botanical theories in maturity, were most certainly sown very early in life by his parents.

1837 The year Ruskin went up to Oxford and a significant purchase was made:[7] the first two volumes of W. Baxter, *British Phænogamous Botany: figures and descriptions of the genera of*

1. Blunt, W. and Stearn, W., *The Art of Botanical Illustration*, new edition, revised and enlarged (Antique Collectors' Club, Woodbridge, in association with The Royal Botanic Gardens, Kew, 1994), p. 272. See also p. 150, this volume.
2. Mabey, R., *Weeds* (Profile Books, London, 2010), p. 148.
3. *Works*, vol. XXXV, *Præterita I*, p.36.
4. *Works*, vol. XXXV, *Præterita I*, p. 59.
5. *Works*, vol. XXV, *Proserpina I & II*.
6. *Works*, vol. XXV, *Proserpina I*, 198.
7. See Ingram, D. S., 'Ruskin's Botanical books: A Survey of Re-ordered and Annotated Second Edition Volumes of *British Phænogamous Botany* (W. Baxter, 1834–43) and *English Botany* (J. E. Smith & J. Sowerby, 1832–1840)', *Ruskin Review and Bulletin* 12 (1) (2016), pp. 18–50; and Ingram, David [S.] (2016) *Ruskin's Botanical Books: Re-ordered and Annotated Editions of* Baxter *and* Sowerby (Guild of St George, York, 2016), p. 81.

Opposite: John Ruskin, Aspen Tree (Populus tremula *L.*), *date unknown, pencil and watercolour*

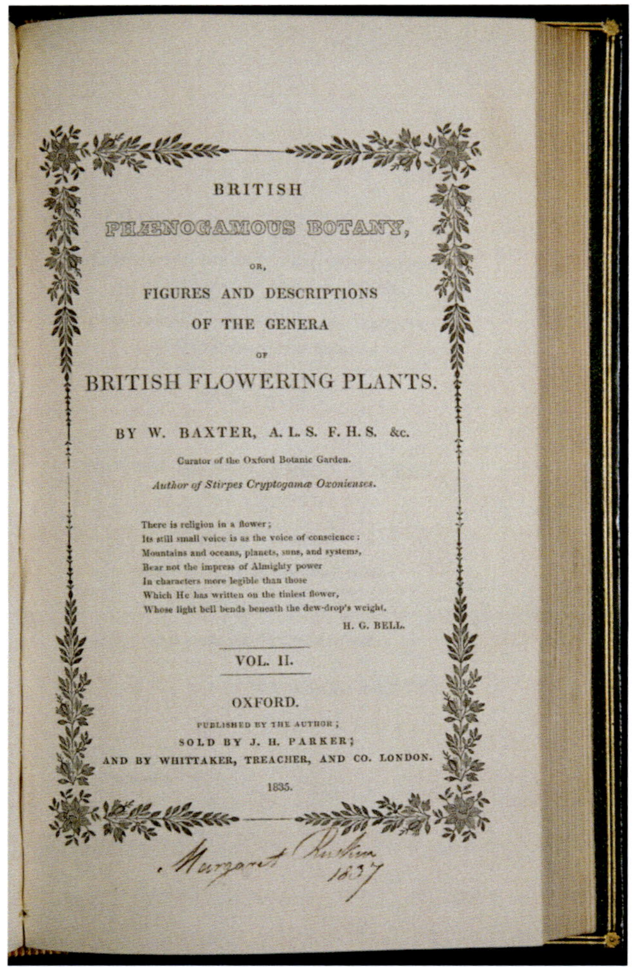
BRITISH

PHÆNOGAMOUS BOTANY,

OR,

FIGURES AND DESCRIPTIONS

OF THE GENERA

OF

BRITISH FLOWERING PLANTS.

BY W. BAXTER, A. L. S. F. H. S. &c.

Curator of the Oxford Botanic Garden.

Author of Stirpes Cryptogamæ Oxonienses.

There is religion in a flower;
Its still small voice is as the voice of conscience:
Mountains and oceans, planets, suns, and systems,
Bear not the impress of Almighty power
In characters more legible than those
Which He has written on the tiniest flower,
Whose light bell bends beneath the dew-drop's weight.
H. G. BELL.

VOL. II.

OXFORD.

PUBLISHED BY THE AUTHOR;

SOLD BY J. H. PARKER;

AND BY WHITTAKER, TREACHER, AND CO. LONDON.

1835.

Margaret Ruskin 1837

Part of the title page of Volume II of W. Baxter, British Phænogamous Botany, *showing Margaret Ruskin's signature, dated 1837*

British flowering plants, 2nd edition (1834–43), with or for his mother. Six volumes of this work were eventually bought and were destined to play, in 1855, an important part in Ruskin's botanical studies.

1842 Notably the year when Ruskin first learned to 'see' plants by drawing them and consolidated that skill during a journey to Chamonix. Writing in *Præterita* he describes walking one day on the road to Norwood in the early months of 1842, and states: 'I noticed a bit of ivy round a thorn stem, which seemed, even to my critical judgment, not ill 'composed'; and proceeded to make a light and shade pencil study of it ... carefully ... When it was done, I saw that I had virtually lost all my time since I was twelve years old, because no one had ever told me to draw what was really there! ... [I] had never seen the beauty of anything, not even of a stone – how much less of a leaf!'[8]

In the early summer of 1842 Ruskin and his parents travelled to Chamonix. Again in *Præterita*, he speaks of drawing an Aspen tree near Fontainebleau:[9] '... and as I drew ... the beautiful lines insisted on being traced,—without weariness. More and more beautiful they became, as each rose out of the rest, and took its place in the air. With wonder increasing every instant,

8. *Works*, vol. XXXV, *Præterita II*, p. 311.
9. *Works*, vol. XXXV, *Præterita II*, pp. 313–314.

John Ruskin, Oxalis (Wood Sorrel), date unknown, pencil and watercolour. Ruskin called Oxalis his 'favourite Chamouni plant' (Diary, February 4th, 1844)

I saw that they 'composed' themselves, by finer laws than any known to men. At last, the tree was there, and everything that I had thought before about trees, nowhere.' Referring back to the earlier experience on the road to Norwood, he declares that the experience with the Aspen was 'an insight into a new sylvan world. … Not sylvan only. The woods, which I had only looked on as wilderness, fulfilled I then saw, in their beauty, the same laws which guided the clouds, divided the light, and balanced the wave … I find no diary whatever of the feelings or discoveries of this year … I did not even draw much,—the things I now saw were beyond drawing—but took to careful botany …'[10]

Ruskin also remembers the botanical significance of the 1842 visit to Chamonix in the Introduction to the first volume of *Proserpina*, writing from Brantwood on March 14th, 1874 that: 'I had begun my studies of Alpine botany … in 1842, by making a careful drawing of wood-sorrel at Chamouni; … and bitterly sorry I am, now, that the work was interrupted. For I drew, then, very delicately; and should have made a pretty book if I could have got peace. Even yet I can manage my point a little, and would far rather be making outlines of flowers than writing; and I meant to have drawn every English and Scottish wild flower …'[11]

10. *Works*, vol. XXXV, *Præterita II*, p. 315.

11. *Works*, vol. XXV, *Proserpina I*, p. 204.

1843: MODERN PAINTERS PART I PUBLISHED The interruption mentioned above by Ruskin was the need to write the first volume of *Modern Painters*, his 'large book' on art,[12] which had its genesis in a proposed pamphlet in defence of J. M. W. Turner against his critics.

The creation of *Modern Painters* was not completed, however, until the publication of the fifth part, in 1860. In part I, Ruskin argued that contemporary painters, notably Turner, who emerged from the tradition of the picturesque, were superior to the old masters in the art of landscape painting. A central tenet of this and subsequent volumes was that the primary concern of art should be 'truth to nature'.

In part II, and of great significance in his development as a botanist and botanical artist, Ruskin analysed in detail the accurate portrayal of plants and trees, placing emphasis on the relationship between their underlying anatomy, physiology and growth, and their ultimate form. To Ruskin it was as important for the landscape artist to represent the natural world as accurately and with as much empathy with the subject matter as it was for the portrait painter to represent a human subject.[13]

Thus Ruskin writes of 'The Leaf' with insight and eloquence, as follows: 'the leaves ... are the feeders of the plant. Their own orderly habit of succession must not interfere with their main business of finding food. Where the sun and air are, the leaf must go, whether it be out of order or not. So, therefore, in any group, the first consideration with the young leaves is ... how to keep out of each other's way, that every one may at once leave its neighbours as much free-air pasture as possible and obtain a relative freedom for itself. This would be a quite simple matter, and produce other simply balanced forms, if each branch ... had nothing to think of but reconcilement of interests among its own leaves. But every branch has others to meet or to cross, sharing with them, in various advantages, what shade, or sun, or rain is to be had. Hence every single leaf-cluster presents the general aspect of a little family, entirely at unity among themselves, but obliged to get their living by various shifts, concessions and infringements of other people in their neighbourhood.'[14]

Although primarily a great work of art criticism, *Modern Painters*, with its emphasis on truth to nature, may also be read, in my view, as a most important contribution to the power of botanical observation and the subsequent recording of what has been observed in words and drawings.

1844: RUSKIN COMPILED HIS *FLORA OF CHAMOUNI*, THE SUBJECT OF THE PRESENT VOLUME It is botanically significant that in the *Flora* he uses entirely conventional scientific names to label his specimens and gives no hint whatsoever, either in the volume itself or in his accompanying Diary entries, of any dissatisfaction with the contents of contemporary botanical books or schemes of plant classification.

1855: A NEW CLASSIFICATION OF PLANTS IS CREATED By this year, Ruskin's botanical studies connected with the writing of *Modern Painters* had clearly led him to become deeply frustrated with the work of contemporary botanical taxonomists, for he wrote a most significant letter to Jane Welsh (Mrs Thomas Carlyle), a friend and regular correspondent, in which he states that: [While writing *Modern Painters*] '... I became dissatisfied with the Linnaean, Jussieuan,[15] and Everybody-elsian arrangement of plants, and have accordingly arranged a system of my own; and unbound my botanical book, and rebound it in brighter green, with all the pages through other, and backside foremost ... and am now printing my new arrangement in a legible manner, on interleaved foolscap. I consider this arrangement one of my great achievements of the year ...'[16] Ruskin returns to this theme five years later (1860), in a letter to a botanical friend,

12. *Works*, vols. III–VII, *Modern Painters*.
13. Ingram, David (S.) & Wildman, S. *Ruskin's Flora: The Botanical Drawings of John Ruskin* (Ruskin Library and Research Centre, Lancaster University, Lancaster, 2011), pp. 14–15.
14. *Works*, vol. VII, *Modern Painters VI*, p. 48.
15. Carl von Linné (Linnaeus) and Antoine-Laurent de Jussieu, early and influential Swedish and French (respectively) plant taxonomists.
16. This letter is quoted in the Introduction to vol. V of *Works*, p. xlix, as being to Mrs Carlyle and is also referred to as being to Mrs Carlyle in vol. XXXVI, p. 183, in the introductory remarks to the letter written from Denmark Hill in 1855. A transcript is also included in *The Correspondence of Thomas Carlyle and John Ruskin*, ed. by George Allan Cate (Stanford University Press, Stanford, 1982), where it is said to be from 'Ruskin to Carlyle', and dated 'ca. October 1855'.

John Ruskin, Rhododendron Leaves, c. 1854–58, pencil, watercolour and bodycolour. This image was reproduced in wood engraving as Fig. 33 of vol. V of Modern Painters *(1860)*

Pauline Trevelyan, as follows: 'I have had … more difficulty than I expected in obtaining as much clue to simple matters in botany, as was required for my account of vegetation form, and in investigating such things as I wanted, for myself … In many respects I am entirely foiled – finding things on which I supposed scientific men were entirely agreed – still in dispute.[17] And finally, in 1865, writing again to Pauline Trevelyan, he expressed a desire to turn botany 'upside down'.[18]

Until 2015, the nature of Ruskin's new classification scheme for plants was not generally known. In that year, however, the Guild of St George purchased from a private owner the six volumes of a second edition of *British Phænogamous Botany* [Flowering Plants], by William Baxter (then Curator, Oxford Botanic Garden), published between 1834 and 1843. Ruskin's copy is listed in Dearden's catalogue of 2012,[19] mentioned by Collingwood in *Ruskin Relics*,[20] and was clearly the botanical book referred to in the letter to Jane Welsh discussed above. I made a detailed study of the volumes immediately after their acquisition by the Guild and this revealed that they represent a significant stage in Ruskin's botanical thinking, one which ultimately found full, idiosyncratic expression in the two volumes of *Proserpina*.[21]

17. *Reflections of a Friendship: John Ruskin's Letters to Pauline Trevelyan, 1848–66*, ed. by Virginia Surtees (George Allen & Unwin Ltd., London, 1979), p. 141.
18. Ibid. p. 248. Ruskin discussed botany with many people in addition to Pauline Trevelyan, especially: his mother, Margaret Ruskin; Jane Welsh Carlyle; Maria La Touche; Susanna (Susie) Beever; William Robinson; and Professor Daniel Oliver, FRS (see Ikin, 2019).
19. Dearden (2012), Cat. no. 170.
20. Collingwood (1903), *Ruskin Relics*, p. 188.
21. *Works*, Vol. XXV, *Proserpina I & II*.

The six volumes of of W. Baxter, British Phænogamous Botany, *1836, as rebound by Ruskin with text and illustrations separated*

The copy of *Baxter*[22] was, as Ruskin had told Jane Carlyle, bound in bright green morocco, with the individual volumes numbered and lettered in gilt on the spine. The contents, if not 'backside foremost', were certainly not as Baxter had ordered them. His original six volumes included all the plates, arranged in the random order in which they had been produced, with each followed by a description of the species illustrated. A series of indexes in the final volume then unified the whole work, both taxonomically and alphabetically. In Ruskin's copy, all the descriptions had been separated from their corresponding plates and bound together in Volumes 1 to 3 (sic), in their original order and with their original page numbers. The coloured plates, however, were bound separately, in Volumes IV to VI (sic), and had been completely re-ordered and re-numbered. It was evident that Volumes I and II of the work had once been owned by Ruskin's mother, Margaret, since early pages had been signed by her in black ink and dated 1837 (see illustration on p. 134).

Ruskin clearly intended that the re-ordered *Baxter* should be put to good use, for he took great care in numbering and cross-referencing the descriptions and plates.

Most significantly, the plates in Volumes IV to VI had been re-arranged by Ruskin into entirely new Classes and Orders with interleaved sheets of pale blue foolscap paper, just as he says in his letter to Jane Carlyle. Although he had retained the Linnæan genus and species names (the binomials) used by Baxter, he

22. Ingram, D. S. (2016a), 'Ruskin's Botanical Books: A Survey of Re-ordered and Annotated Editions of *British Phænogamous Botany* (W. Baxter, 1834–43) and *English Botany* (J. E. Smith & J. Sowerby, 1832–1840)' *Ruskin Review and Bulletin* 12 (1), pp. 18–50; and Ingram, David [S.] (2016b), *Ruskin's Botanical Books: Reordered and Annotated Editions of* Baxter *and* Sowerby (Guild of St George, York). Finally, I have included in the present account, in some cases near verbatim, edited material based on the following summary paper: Ingram, David [S.], (2016c) 'John Ruskin's botanical books: new discoveries and work in progress', *The Companion*, Journal of the Guild of St George, 16, pp. 7–11.

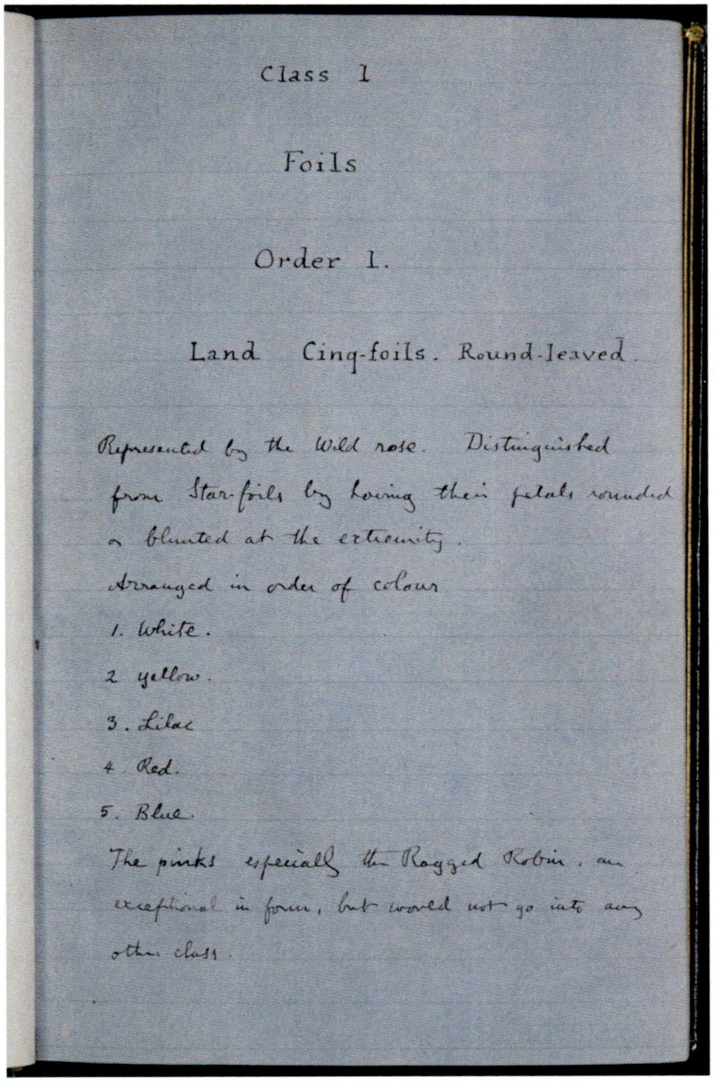
Class I

Foils

Order I.

Land Cinq-foils. Round-leaved.

Represented by the Wild rose. Distinguished from Star-foils by having their petals rounded or blunted at the extremity.

Arranged in order of colour

1. White.

2 yellow.

3. Lilac

4. Red.

5. Blue.

The pinks especially the Ragged Robin, are exceptional in form, but would not go into any other class.

Left: A blue, interleaved page in the re-ordered and annotated Baxter inscribed by John Ruskin with details of his novel plant 'Order', 'Land Cinq-foils'. Right: A plate showing a typical 'Land Cinq-foil', Trailing Dog Rose (Rosa arvensis), *in Volume IV of Ruskin's re-ordered Baxter*

completely ignored all existing taxonomies for the higher levels of classification and grouped the plates into five new 'Classes' of his own devising: I. Foils (flowers with un-joined petals); II. Bells (flowers bell-like); III Hoods (flowers hooded); IV Grasses (true grasses and plants that look like grasses); and V. Waywards (plants which he could not fit into the previous four classes). Each of these Classes was then subdivided into 'Orders', on the basis of a variety of unrelated, idiosyncratic and subjective criteria including, variously: petal number, shape and colour; plant size or form; habitat (dry or wet, land or water); flower form or similarity to the apparel of particular people; inflorescence structure; uses, especially as medicines or food; undesirable properties (e.g. poisonous, weedy, spiny); and supposed representation of particular human character traits or conditions (e.g. chattiness, spitefulness, gender or old age). Explicitly male (stamens) and female (pistil: stigma, style and ovary) characters, which were used in scientific classifications at the time, as today, were completely ignored.

Most of the characteristics chosen by Ruskin for his classification were too variable, too subjective and therefore too unreliable to be used as the basis of a scientific classification that takes proper account of biological relatedness among families, genera and species. Nevertheless, the scheme does provide a witty and picturesque set of criteria that a non-specialist trying to put a name to a plant specimen might find extremely useful.[23] The following transcription of the Orders included in the Class 'Hoods' provides a detailed example of the supposed plant characteristics that Ruskin used to form his new classification scheme:

CLASS 3. HOODS

ORDERS:

1. MONK'S HOODS. Apt to be dangerous, and connected with Snaps of Dragons, and Gloves of Foxes. Type, the Arum; when...[unreadable word]...and well hooded as the Arum, very beautiful. [e.g. Monk's-hood and Purple Foxglove.]
2. KNIGHT'S HOODS. Known by the attached Spurs. [e.g. Columbine.]
3. SAILOR'S HOODS. Arranged in clusters on Masts, above leaves set like Mast heads on "Tops". [e.g. White dead-nettle.]
4. MONKEY'S HOODS. Having a strange gift of Imitation. [Mainly Orchid family; e.g. Bee Orchid and Monkey Orchid.]
5. CLUSTERED HOODS. [e.g. Blue-bottle; now called Cornflower.]
6. BRANCHING HOODS. [e.g. Common Marjoram.]
7. OLD LADIES' HOODS. Generally stooping or creeping; and very good for making tea, or medicinal draughts. [e.g. Wild Thyme.]
8. YOUNG LADIES' HOODS. Generally pleasant to behold, and serviceable in households [e.g. Dyer's Green-weed]; but apt to be very troublesome in the form of Tares [e.g. Prickly Rest-harrow]. Sometimes showing inclinations towards gay bonnets. [e.g. Everlasting Pea.]

The pages of Baxter were also scattered with cross-references and marginal and text notes written in black ink in Ruskin's unmistakable hand. The most common, over forty, were cross-references to 'F', an abbreviation for *Flora Danica*.[24] Since Ruskin did not acquire his copy of *Flora Danica* until 1866,[25] it is presumed that these cross-references were inserted during or after that date, more than ten years after the volumes of *Baxter* were re-ordered and rebound, perhaps during a period of botanical activity following his acquisition of such an important and beautiful work. That Ruskin greatly respected the authority of *Flora Danica* is implied by the association of abbreviations of the word 'confirmed' with many of his numerous cross-references to the work.

References to the taxonomic works of the many eminent late eighteenth and early nineteenth century plant taxonomists were rare in the annotations and cross-references in the re-ordered *Baxter* and in the new classification itself. However, just as Collingwood observes concerning Ruskin's botanical books, the re-classified *Baxter* reflects 'his purely artistic and unscientific interest in natural history'.[26] Ruskin's new scheme for classifying the plants in *Baxter*, employing criteria important to a nineteenth century artist rather than a scientist, is a pointer to the taxonomic excesses that were to come later in *Proserpina*.

1868 Ruskin wrote *Moss*, dated Denmark Hill, 3rd November 1868, a paper later included as Chapter 1 of *Proserpina*. His approach is characteristically, by turns, idiosyncratic, poetic, witty and perceptive. His stream of consciousness style of writing here, so typical of all Ruskin's botanical writing—although self-indulgent in one sense—is very revealing about his way of working. We, the readers, sit with him at the desk in his study, then walk with him into the garden, returning to the desk with a mossy brick; we then observe with him the form of the moss leaves through his lens 'of no great power',[27] we scan with him his bookshelves and search through favoured volumes for enlightenment and inspiration; and we share his delight in the beauty of nature, his frustration with, as he sees it, the deficiencies of scientific writers, his moments of discovery, his pleasures, his disappointments and his humour. I recommend *Moss* as a most entertaining piece to read as an introduction to Ruskin's

23. Indeed, some modern field guides group plant illustrations according to flower colour, e.g. Hoppe, Ansgar, *A Field Guide to the Flowers of the Alps* (Pelagic Publishing, Exeter, 2013).

24. Oeder, G. C. et al., *Icones Plantarum sponte nascentium in regnis Daniæ et Norvegiæ*, 10 vols (Copenhagen, 1766–1823) and *Supplement* of 9 vols (1829–65).

25. Dearden (2012), catalogue no. 1907.

26. Collingwood (1903), *Ruskin Relics*, p. 188; see note 29.

27. Ruskin is well known to have abhorred the use of scientific instruments of high magnification.

Brantwood today from the eastern shore of Coniston Water

botanical writing, not least the extended account of the false logic which leads him to conclude that there is a close taxonomic link between mosses and members of the family Bromeliaceae. This family includes pineapples which, according to Ruskin: 'really seem to be a kind of moss on a vast scale.'[28]

1871 Ruskin purchased the house at Brantwood and its surrounding land, making it his home from 1872 until his death in 1900.[29] In a letter he described his new acquisition, which was situated on the eastern shore of Coniston Water in the English Lake District, as 'five acres of rock and moor; a streamlet, and I think … the finest view I know in Cumberland or Lancashire';[30] a modest description of what was to become a consuming passion. It provided a private world where he could relax after arduous travels or long hours of writing and lecturing. It was also a place where he could find the peace to work out, sometimes quite literally on the ground, his intellectual, social and practical ideas, including those relating to plants. However, when his mental health began to deteriorate, it became a thorn in his side, a landscape of memory reminding him of his parents, the unsuccessful pursuit of love, especially of Rose La Touche, and of failed projects.[31]

28. Ruskin, J., *Proserpina*, vol. I (George Allen, Sunnyside, Orpington, Kent, 1882), pp. 13–17; (*Works*, vol. XXV, pp. 207–210).
29. In writing this short summary of the relevance of the gardens at Brantwood to Ruskin's development as a botanist, I have drawn on three publications, as follows: W. G. Collingwood, 'Ruskin's Gardening', in *Ruskin Relics* (Isbister & Co. Ltd., London, 1903), pp. 11–44 (written shortly after Ruskin's death by a man who knew and admired him); John Illingworth, 'Ruskin and Gardening', *Garden History*, journal of the Garden History Society, 22, (1994), pp. 218–233 (a most perceptive, scholarly study); and David (S.) Ingram, *The Gardens at Brantwood: evolution of John Ruskin's Lakeland paradise* (Pallas Athene, London, 2014), an illustrated account, for the general reader, of the history and present state of the Brantwood gardens of Ruskin and Joan Severn, re-imagined and augmented by Sally Beamish, Howard Hull and other Brantwood staff.
30. Letter to Charles Eliot Norton, September 15th, 1861, *Works*, vol. XXXVII, p. 35.
31. Ikin, Caroline, *Reading Ruskin in the garden: the designed landscape at Brantwood (1871–1900)*, a thesis submitted in partial fulfilment of the requirements of Manchester Metropolitan University for the degree of Doctor of Philosophy, 2019.
32. *Works*, vol. XXXV, *Præterita I*, p.59.

John Ruskin, A single petal from a Rhododendron flower at Brantwood, pencil. Drawn 'to show principle of order in pardonable disorder'

A significant feature of the land at Brantwood so far as Ruskin's development as a 'botanist' is concerned, is, I believe, the diversity of its landscapes, its habitats and in consequence its plant species. It became a place where Ruskin's time could once again, perhaps for the first time since childhood, be passed: 'chiefly in the … close watching of the ways of plants … staring at them, or into them … in admiring wonder … [pulling] every flower to pieces till I knew all that could be seen of it' (see CHILDHOOD, p. 133 above, for context).[32]

After purchasing the house, he almost immediately set about managing the landscape in his own way, a new way that would ultimately have a significant impact in terms of botanical diversity. The intervention began modestly. His first visit to the house began on 12th September 1871 and then, three days later, after taking stock of his acquisition and metaphorically rolling up his sleeves, he began weeding, writing during the evening:[33] 'First day's work clearing garden. See weeds burnt in twilight.' When he became established as a resident in his new domain, from 1872 onwards, there began a long period of more serious gardening activities, such as clearing the bed of the beck, woodcutting and undertaking Herculean excavations of enormous quantities of soil and rocks to create new gardens, garden structures and extensions to the house. Gradually the landscape was transformed, and herein lies a paradox: although Ruskin loved 'wild' landscapes and 'wild' flowers, as a gardener he was an unrepentant interventionist.

33. Diary, September 15th 1871. This and all subsequent 1844 Diary summaries and quotations are based on the manuscript Diary for 1844 held in the Whitehouse Collection, The Ruskin, Lancaster University, (RF MS 6) and on Evans, J. & Whitehouse, J. H. (1956), *The Diaries of John Ruskin, Vol. 1, 1835–1847* (Clarendon Press, Oxford), pp. 275-296.
34. Collingwood (1903), p. 31.

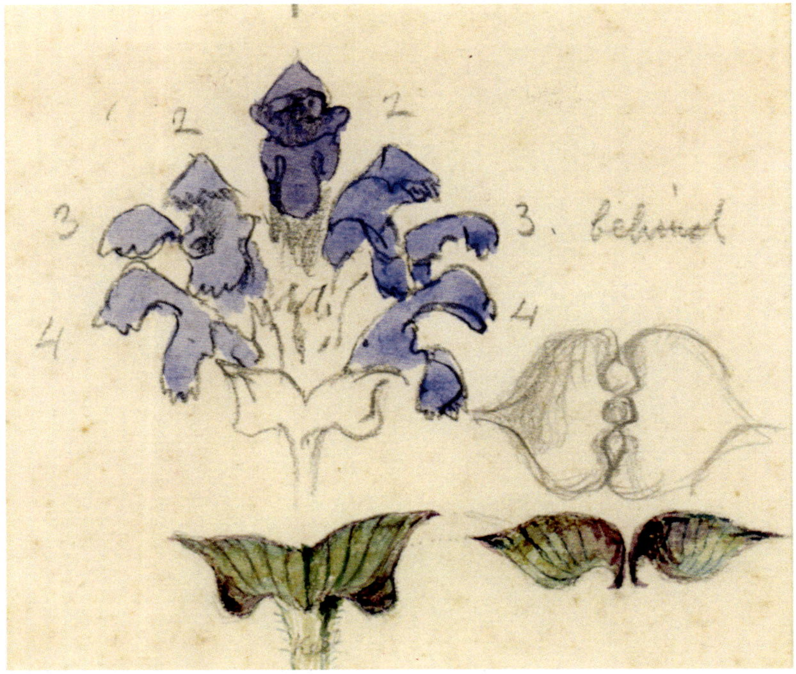

John Ruskin, A dissected inflorescence of Common Self-heal (Prunella vulgaris), *pencil and watercolour*

W. G. Collingwood illustrates this paradox perfectly, writing: 'though he wrote about botany, and tried to be his own Linnæus, and though he loved well enough to see flowers (especially wild ones) on his table and outside his window, yet in his practical gardening was quite the landscapist. He liked making paths and contriving pretty nooks, building steps and bridges, laying out beds, woodcutting and so forth; but I never remember him potting and grafting and layering and budding; and as to the rarity of any plants in his garden, I believe he took far more pleasure in the wood-anemone— Silvia, he called it—than in anything buyable from the nurseryman's catalogue.'[34] In Ruskin's defence, however, it must be emphasized that his soft landscaping was contrived to look as natural as possible and his hard landscaping, although of stone or wood, was ephemeral, such that it could, and indeed often did, eventually crumble away. Moreover, the results of his labours probably resulted in a greater diversity of habitats and therefore more wild plant species and traditional garden plants than were there before his arrival.

Ruskin's first major act of intervention in the Brantwood landscape concerned the coppicing of the woodland. Thus Collingwood writes: '... he would have his coppice cut no more. He let it grow, only taking off the weaker shoots and dead wood. It spindled up to great tall stems, slender and sinuous, promising no timber, and passed the age for all commercial use or time-honoured wont. Neighbours shook their heads, but they did not know the pictures of Botticelli, and Ruskin had made his coppice into an early Italian altar-piece.'[35] But since the trees were living and therefore continued to grow: 'to keep his forest at this delightful point of mystery, his billhook and gloves were always lying on the hall table, and after the morning's writing he would go up to the Brant (steep) Wood and chop for half an hour before luncheon. It was not the heroic axe-work of Mr. Gladstone, but such pruning as a Garden of Eden required to dress it and keep it.'[36]

Discontinuing the practice of full coppicing in this way must have resulted in the woodland being held at a point somewhat equidistant between the full light at the beginning of the coppice

35. Collingwood (1903), p. 35.
36. Collingwood (1903), p. 36.

The Professor's Garden, Brantwood

cycle and the deep shade at its conclusion. This meant that only the woodland plants that could thrive in dappled shade would remain. The light-loving plants would not have been lost, however, for coppicing would have continued in the woodlands on the land adjacent to Brantwood. Moreover, Ruskin's further landscaping activities, the creation of a series of specialized gardens, embedded within the woodland and interlinked by or embracing a labyrinth of mossy paths, irregular stone steps, fences, low walls and terraces, were destined to create more and more habitats where light-loving plants might flourish in abundance. These gardens together constitute what I like to think of as 'gardens in a wooded landscape' rather than a coherent landscape garden.

So the newly turned earth of, for example, the beds of what is now called the Professor's Garden, the terraces of Zig-zaggy and the Moorland Garden would have provided a rich substrate for light-loving annual and perennial wild plants to flourish. The moist shade of his Tennis Ground (now called the Painter's Glade) would have furnished the perfect habitat for liverworts, mosses and ferns, as it still does. The steps, paths, walls and cliffs of all the gardens would have doubled as rock gardens to supplement that of his Precipice Garden, an alpine garden in miniature. The digging of a reflecting pond in 1877–8 and reservoirs in the Moorland Garden would have resulted in relatively deep water with shallow, boggy margins, where aquatic and marsh plants might flourish, to augment the existing flora of ferns, primulas and other moisture and shade-loving species at the side of the tumbling woodland beck. And last, the massive groundworks associated with the construction of the Tennis Ground, Zig-zaggy and the Moorland Garden, together with the laying out of an additional kitchen garden, the construction of an underground ice-house and water cistern, and the digging of the foundations for extensions to the house and outbuildings, must have involved the excavation of immense quantities of rock and soil. This, presumably, would have exposed to sunlight the dormant seeds of wild annual flowers, thereby triggering germination. These species, their diversity enhanced by wild species already present in the woodland soil, and annuals and perennials growing from seeds deposited on the bare earth by wind, birds and animals, must have sprung into life in the following seasons. Before long the new soil would then also have been slowly colonized by a succession of grasses, perennial herbs, woody shrubs and

The beck, Brantwood

tree species: without doubt Ruskin's landscape interventions at Brantwood would have provided him with a rich harvest of plant specimens for his botanical studies as he worked on *Proserpina*.

1875: PUBLICATION OF PROSERPINA, PART I The idiosyncratic climax of Ruskin's ideas concerning botanical naming and classification began with what would become his major, two-volume work on the study of plants: *Proserpina—Studies of Wayside Flowers, While the Air Was Yet Pure Among the Alps, And in Scotland And England Which My Father Knew*.

A book of botany had been in Ruskin's mind for some time, probably even before 1868 when he wrote *Moss*. Interestingly, in the context of the present volume, an early working title suggested that it would have a clear Alpine and Arctic focus, for in a letter to Charles Eliot Norton from Denmark Hill, dated November 17th, 1869, he writes: 'I write every day, if possible, a little of my botany … to be called *Cora Nivalis*, "Snowy Proserpine": an introduction for young people to the study of Alpine and Arctic wild flowers.'[37] This emphasis clearly changed to the wayside flowers of the Alps, Scotland and England during subsequent writing.

The name finally chosen by Ruskin for his botanical work—*Proserpina*—was that of a Roman deity, the daughter of Ceres, goddess of agriculture and crops, and Jupiter, the god of sky

37. *Works*, vol. XXXVI, pp. 597–8; Hilton, T., *John Ruskin: the Later Years* (Yale University Press, New Haven and London, 2000), p. 310.

John Ruskin, Study of Roses, from the dress of Spring in Botticelli's Primavera, *1874, pencil, ink, watercolour and body colour*

and thunder and king of the gods. Her story forms the basis of a myth of Springtime, and her name, appropriately, derives from the Latin *prospere*, to emerge, as in the germination of a grain of wheat. Following publication of the first part of *Proserpina*, in 1875, nine further parts were published, the last not appearing until 1886. All were illustrated with engravings by Ruskin's former pupil, George Allen, and woodcuts by Arthur Burgess, based on Ruskin's own paintings and drawings, many merely sketches. The ten parts of *Proserpina*, together with an introduction and additional botanical writing by Ruskin, were later brought together by George Allen as two volumes, the first being published in 1882 and the second in 1888; further unpublished material subsequently appeared in the collected *Works of John Ruskin*.[38] Much of the material comprising *Proserpina* was written at Brantwood and, as has already been suggested, many of the ideas and illustrations included in the work no doubt had their origins among the plants growing in the gardens there or in the surrounding fells and dales.

Ruskin introduces the first part of *Proserpina* with an account of why he undertook the work.[39] First he speaks of p. 318 of the 1795 volume of Curtis's *Botanical Magazine*, which describes:

> what I suppose to be a favourite flower of mine, called in Curtis, The St Bruno Lily. [This] page of text [which includes six alternative scientific names for St Bruno's Lily] … is so characteristic of botanical books, and botanical science, not to say all science as hitherto taught for the blessing of mankind; and of the difficulties thereby accompanying its communication …

After roundly condemning p. 318, and with it all botanical science, and blaming his possession of Curtis for his botanical ignorance (although excusing his parents for buying it for him; see CHILDHOOD, above) he explains that *Proserpina* was written:

> to put … some elements of the science of botany into a form more tenable by the ordinary human and childish faculties; …—to make the paths of approach to it more pleasant.

There then follow two volumes of what may be described as a misconceived obsession with the creation of an accessible, counter-Linnæan taxonomy of plants, based on æsthetic, spiritual and human characteristics rather than a scientific analysis of the reproductive structures and principles used by Linnæus and his plant taxonomic successors, a process that Ruskin had begun

38. *Works*, vol. XXV, *Proserpina I*, pp. 549–569.

39. *Works*, vol. XXV, *Proserpina I*, pp. 197–200.

with the re-ordering of Baxter in 1865. Throughout *Proserpina* plants are given human minds, personalities and traits, and the Divine purpose is frequently referred to. It is, nevertheless, a fascinating book, containing passages of great eloquence and beauty, as when Ruskin describes a poppy flower:

> It is an intensely simple … flower. All silk and flame: a scarlet cup, perfect-edged all round, seen among the wild grass far away, like a burning coal fallen from Heaven's altars. You cannot have a more complete, a more stainless, type of flower absolute; inside and outside; all flower.[40]

Elsewhere, as in *Modern Painters*, there are many passages of botanical insight, as when he describes a weed, for example, possibly through personal experience at Brantwood, as: 'A vegetable which has an innate disposition to get into the wrong place … It is not its being venomous, or ugly, but its being impertinent – thrusting itself where it has no business, and hinders other people's business – that makes a weed of it.'[41] With similar insight he describes a leaf as follows: 'This *thin* film I said. That is the essential character of the leaf; to be thin, – widely spread out in proportion to its mass. It is the opening of the substance of the earth to the air, which is the giver of life.'[42] A large surface area to volume ratio is indeed the key to the success of the leaf as an organ of photosynthesis,[43] providing the maximum surface area possible for the collection of light energy to fuel the process.

Ruskin's new classification is, however, highly controversial, although fascinating reading nonetheless. As an example, it is necessary first to recall that in the re-ordered Baxter (see above), where one of Ruskin's new classes is 'Foils', this group is further divided into ten 'Orders' of foils, each illustrated with examples, which include the original Linnæan genus and species names. In *Proserpina*, however, Ruskin goes one step further and devises new genus and sometimes species names. Thus, with regard to 'Cinq-foils [flowers having five petals]' he writes:[44]

> My younger readers will find it easy to learn, and convenient to remember, for a beginning of their science, the names of twelve great families of cinquefoiled flowers, of which the first group of three, is for the most part golden, the second, blue, the third, purple, and the fourth, red. And their names, by simple lips, can be pleasantly said, or sung, in this order, the two first only being a little difficult to get over.
>
> 1. Roof-foil, Rock-foil, Primrose.
> 2. Lucy, Blue-bell, Bindweed.
> 3. Pea, Pansy, Daisy.
> 4. Pink, Peach, Rose.

A footnote explains that if the florets of the Daisy are examined closely they will be seen to be 'cinquefoils'. Ruskin then goes on to give his own 'approved' scientific equivalents:

> Which even in their Latin magniloquence will not be too terrible, namely:—
>
> 1. Stella, Francesca, Primula.
> 2. Lucia, Campanula, Convoluta.
> 3. Alata, Viola, Margarita.
> 4. Clarissa, Persica, Rosa.

Only Primula, Campanula, Viola and Rosa were accepted scientific genus names, the other eight names in the list being devised by Ruskin.

There is insufficient space here to give Ruskin's logic for making all of these changes, but a single example, 'Francesca', gives the flavour, as follows:

> The second tribe, (at present Saxifraga[45]) growing for the most part wild on rock, may, I trust, even in Protestant botany, be renamed Francesca, after St. Francis of Assisi; not only for its modesty, and love of mountain gravel, and poverty of colour and leaf; but also because the chief element of its decoration, seen close, will be found in its spots, or stigmata.[46]

It is important to draw attention to Ruskin's desire, throughout his new classification scheme and the descriptions

40. *Works*, vol. XXV, *Proserpina I*, pp. 253–254.
41. *Works*, vol. XXV, *Proserpina I*, pp. 283–284.
42. *Works*, vol. XXV, *Proserpina I*, p. 230.
43. Ingram, David S., Vince-Prue, D. & Gregory, P. J., *Science and the Garden*, (Wiley-Blackwell, Oxford, 3rd edition, 2016), p. 17.
44. *Works*, XXV, *Proserpina I*, pp. 314–315.
45. An Alpine genus represented in the *Flora of Chamouni 1844*.
46. *Works*, vol. XXV, *Proserpina I*, p. 315.

of plants within it, to expunge words which he describes in the Introduction to *Proserpina* as being: 'founded on some unclean or debasing association, so that to interpret them is to defile the reader's mind.'[47] An example of this is his problem in finding a title for Plate VII, an illustration of a plant which he calls 'Contorta purpurea', 'Purple Wreath-wort', in fact a representation of an orchid, or *Orchis* species. The plants of this genus frequently have bulbous roots, hence the scientific name based on the Classical Greek word *orchis*, meaning a testicle. Ruskin writes in *Proserpina*:

> It is one of the flowers whose names I think good to change; but I look carefully through the existing titles belonging to it and its fellows, that I may keep all I expediently can. I find, in the first place, that Linnæus called one group of its relations, Ophryds, from Ophrys,—Greek for the eyebrow,—on account of their resemblance to the brow of an animal frowning, or to the overshadowing casque of a helmet. I perceive this to be really a very general aspect of the flower; and therefore, no less than in respect to Linnæus, I adopt his for the total name of the order, and call them 'Ophrydæ,' or, shortly, 'Ophryds'.[48]

Wilfrid Blunt and William Stearn, legendary writers on botanical art and plant taxonomy respectively, as noted at the beginning of this chapter, sum all this up perfectly in their 'Digression' on Ruskin in the *The Art of Botanical Illustration*.[49] Firstly, they describe *Proserpina*, despite all its imperfections, as 'the most stimulating book ever written about flowers.' And then, a few lines later, they declare that: 'To non-botanists there was much sense in … [Ruskin's] criticism [of professional botanists] and had he stopped here all would have been well; but he was rash enough to enter the lists himself with a new system of classification which, though poetic enough, was too laughable to be taken seriously by men of science.'

It is my conviction that Ruskin had the intellectual ability to make an enduring contribution to plant taxonomic study. That he did not do so in the writing of *Proserpina* was the result of his misguided and obsessive intellectual and emotional prejudices and inflexibility, causing him to reject the classification schemes for plants devised and accepted by professional botanists both before and during his own lifetime, and thereby rendering himself unable to build on their strengths.

DURING OR AFTER THE WRITING OF PROSERPINA At an unknown date after 1874, Ruskin apparently acquired a most conventional taxonomic text, a seventh edition of *The London Catalogue Of British Plants*, a booklet edited by H. C. Watson, a respected and influential taxonomic editor. Published in 1874 by the London Botanical Exchange Club, Watson's *Catalogue* was intended as a standard to be used by botanists, particularly amateurs, making and classifying their own herbaria and collections, and exchanging specimens with fellow enthusiasts. At some point Ruskin then used this as a basis when he dismantled, re-ordered and rebound the first seven volumes[50] of a second edition of Sowerby's *English Botany*.[51]

Into Volume I he bound *The London Catalogue*, followed by the descriptions of the genera and species of all the flowering plants in the order in which they were originally printed, but with all the plates removed. Each page of the descriptions is numbered, in pencil, in Ruskin's hand.

Into Volumes II–VII he bound all the plates of the flowering plants described in Volume I, but rearranged in the order in which the species are listed in *The London Catalogue*. Each plate is numbered, again in pencil in Ruskin's hand, with the corresponding reference in *The London Catalogue*.

At the end of Volume I, in black ink and in his best handwriting,[52] he added an alphabetical Index listing the genera of the plants illustrated in Volumes II–VII, the volume number for the plates of each genus, the plate number of the first species of that genus illustrated and the page number of the genus description in Volume I.

47. *Works*, vol. XXV, *Proserpina I*, p. 201.
48. *Works*, vol. XXV , *Proserpina I*, p. 341.
49. Blunt, W. & and Stearn, W., *The Art of Botanical Illustration*, (1994), p. 272.
50. Which deal with flowering plants.
51. J. E. Smith & J. Sowerby's *English Botany* (1832–40), usually known as *Sowerby*. See for details: Ingram, D. S., *Ruskin Review and Bulletin* 12 (1) (2016), pp. 18–50; and Ingram, David [S.], *Ruskin's Botanical Books: Re-ordered and Annotated Editions of* Baxter *and* Sowerby (2016).
52. Stephen Wildman, personal communication.

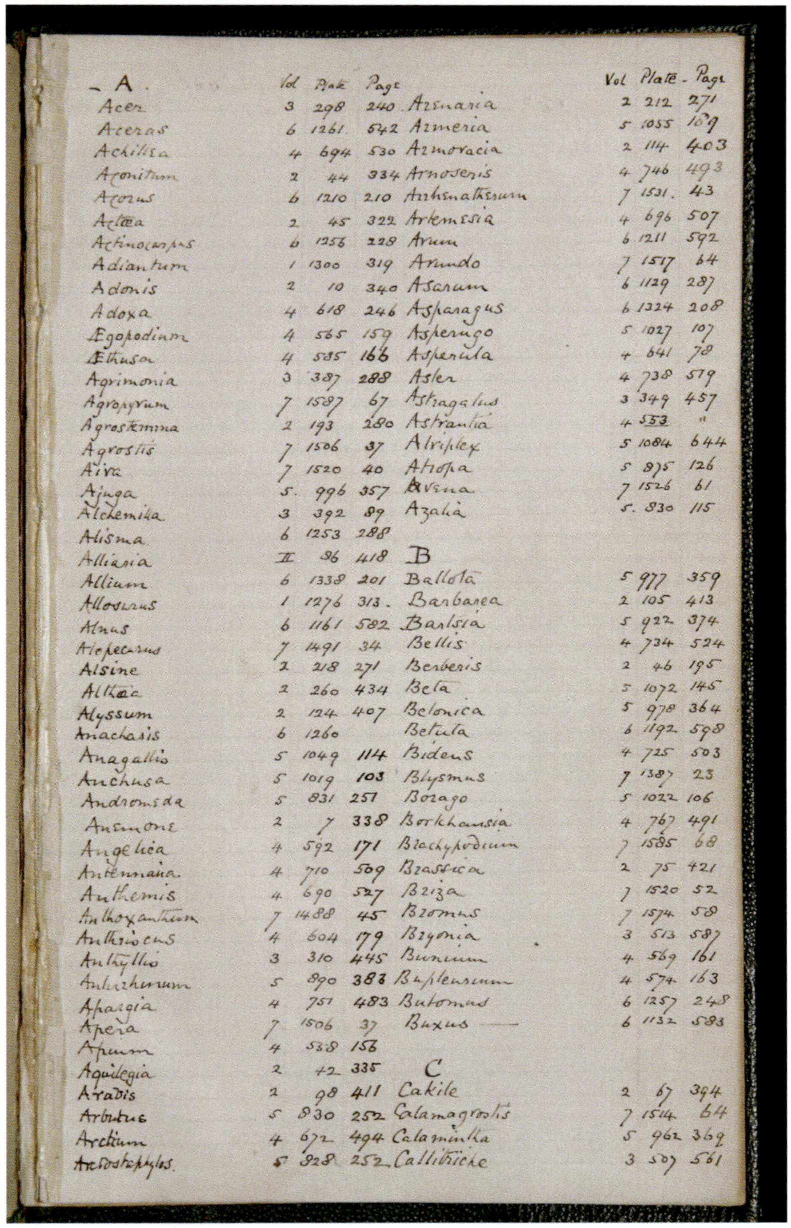

The initial page of Ruskin's index to the re-ordered and rebound volumes of Sowerby

Why and precisely when after 1874 Ruskin was prepared to devote considerable time and energy to this massive undertaking is not known. It is possible that while still writing *Proserpina* he wished to have a yardstick against which he might measure his own novel plant classification scheme. Or maybe, at a much later date, his creative and critical botanical energies having been exhausted by the writing of *Proserpina* and by illness, he was finally willing to accept without challenge Watson's pragmatic and conventional scheme. Whatever his reasons for the re-ordering, however, this project provides an intriguing conclusion to our knowledge of Ruskin's botanical studies.

GAZETTEER OF CHAMONIX AND ENVIRONS

with alternative and contemporary spellings

AIGUILLES DE CHAMONIX Named mountain peaks, north of Mont Blanc: those best known to Ruskin, from north to south, are the Verte, Dru, Charmoz, Blaitière, Plan and Midi. Punning on the literal name, 'needles', his father wrote in 1837 that John knew 'the shape of every needle round Mont Blanc, and could not tell you now where Threadneedle Street is'.

AIGUILLES ROUGES Smaller peaks on the west side of the valley.

ARGENTIÈRE Small town five miles north of Chamonix, on the road ascending the Col de la Forclaz into Switzerland. The bottom of its eponymous glacier is visible from the village.

ARVEYRON [Arveiron, Arveron] River flowing into the ARVE at Chamonix. Its source, formerly issuing from a cave in the Glacier des Bois below the Mer de Glace, was painted by early English artist visitors, including Francis Towne in 1781 and J. M. W. Turner in 1802.

BÉRARD The Cascade de Bérard (waterfall) is a popular beauty spot near the village of LE BUET, north of Argentière.

GLACIER DES BOIS Named after the hamlet of Les Bois, and forming the lower half of the Mer de Glace, the subject of a Ruskin vignette published in 1844. It has all but disappeared since the mid-nineteenth century.

GLACIER DES BOSSONS Named after the hamlet of Les Bossons just south of Chamonix, one of the largest Alpine glaciers, fed from the icefields of Mont Blanc. In the seventeenth century it reached as far as the town, but by the end of the eighteenth had already receded considerably, and continues to melt. Ruskin was one of many regular visitors to notice its gradual retreat.

LE BRÉVENT [Breven] Mountain overlooking Chamonix to the west, offering simple walks to attain fine views of the Aiguilles de Chamonix and Mont Blanc. The commemorative sculpted roundel on the Pierre à Ruskin, unveiled in 1925, is close to the Brévent cable car station.

CHAMONIX [Chamouni, Chamounix] Properly, Chamonix-Mont-Blanc. Town in the Haute-Savoie department of the Auvergne-Rhône-Alpes region of France, 50 miles (81.6 kilometres) from Geneva and 27 miles (44 kilometres) from Martigny. Described in *Bradshaw's Pedestrian Route-Book* (1868) as 'probably the most frequented spot in the High Alps'.

CHARMOZ Northern part of the Aiguilles de Chamonix, above the Mer de Glace. Now called the Grands Charmoz, its lower and more accessible part was known to Ruskin as the PETIT CHARMOZ.

COL DE BALME Mountain pass between the Tête de Balme and Les Grands Otanes, on the border of France and Switzerland north of Argentière.

LA FLÉGÈRE Sits at the base of the Aiguilles Rouges between Chamonix and Argentière. 'It gives the opportunity of seeing the whole line of Mont Blanc ... from one of the best positions which can be obtained – very near and very lofty at the same time.' (Francis Chenevix Trench, *A Walk round Mont Blanc*, 1847)

MER DE GLACE Perhaps the most famous glacier in the Alps, east of Chamonix, formed from the icefields of the Vallée Blanche and technically the Glacier de Montenvers. A tourist attraction since the eighteenth century: William Coxe's 1777 descriptive comparison with 'waves instantaneously frozen in the midst of a violent storm' may have contributed to its popular name, the Sea of Ice.

MONT BLANC At 15,777 feet (4,808 metres), the highest mountain in the Alps. First climbed by Jacques Balmat in 1786, and a favourite of the Alpine Club. Ruskin first saw the mountain in 1833, and nearly fifty years later it remained 'most precious' to him.

MONTENVERS [Montanvers, Montanvert] A steep path from Chamonix to Montenvers gives access to the Mer de Glace. A refuge hut dating back to 1779 was rebuilt in stone in 1840 and replaced in 1879; its present incarnation is the Refuge du Montenvers hotel.

CASCADE DES PÈLERINS (Pilgrims) Now known as the Cascade du Dard, a dramatic waterfall emanating from the Aiguille Blaitière and easily reached from Chamonix.

SALLANCHES [Sallenches] Lying between two branches of the river Arve, affording the first views of Mont Blanc for those travelling south from Bonneville and Cluses towards Chamonix, as J. M. W. Turner did in 1802 and Ruskin in 1835. The town was rebuilt after a great fire in 1840.

ST MARTIN Across the St. Eloi bridge from Sallanches, a small village boasting the popular Hotel du Mont Blanc, which provided the title for Chapter XI of Ruskin's *Præterita* – 'to me, certainly, of all my inn homes, the most eventful, pathetic, and sacred'.

GLACIER DE TACONNAZ Adjoins the Glacier des Bossons, near Les Houches, separated by the Montagne de la Côte.

TAPIA A large slope of moraine and glacial rubble stretching below the Aiguilles de Chamonix between Montenvers and the Plan de l'Aiguille.

TÊTE NOIRE With a famous gorge, now lies just inside the Swiss border on the road from Chamonix to Martigny.

DETAILS OF ILLUSTRATIONS

Frontispiece: John Ruskin, *Aiguille Charmoz, from window of the Hotel de l'Union, Chamonix*, 1849, pencil, ink, watercolour, 30 x 40 cm, Whitehouse Collection, The Ruskin, Lancaster University (1996P0892)

Opposite table of contents: George Richmond, *John Ruskin (The Author of Modern Painters)*, 1843, photogravure after watercolour now destroyed

Opposite dedication: Ruskin memorial stone, Chamonix, incorporating medallion portrait by Michel de Tarnowski and inaugurated 19 July 1925

Page 6: John Ruskin, *Glacier des Bossons, Chamonix*, 1849, pencil, ink, ink wash and bodycolour, 33.8 x 47.7 cm, Ashmolean Museum, Oxford (WA.RS.REF.0091)

Page 8: Carle Vernet, *Les Voyageurs Anglais*, *c.* 1830, lithograph, 37 x 46 cm

Page 9: Jean Dubois, *Hôtel de L'Union, Chamonix,* c. 1820, coloured aquatint; image courtesy of Professor David Hill

Page 10: John Ruskin, *Mont Blanc and the Lake of Geneva from the Jura*, 1835, pencil, ink and bodycolour, 25.5 x 35 cm, private collection

Page 12: J. M. W. Turner, *Mer de Glace*, from *Liber Studiorum*, 1812, etching and mezzotint, 17.8 x 25.7 cm

Page 13: John Ruskin, *Aiguilles of Chamonix from Les Houches*, 1842, pencil, watercolour and bodycolour, 33 x 46.2 cm, private collection

Page 15: Karl-Rudolph Weibel-Comtesse, *Joseph Couttet, aged 55, 1847*, watercolour and bodycolour, Département de la Haute-Savoie collection

Page 17: John Ruskin, *Aiguilles of Chamonix from the foot of La Flégère*, 1849, pencil and watercolour, 37.6 x 47 cm, Birmingham Museums Trust (1907P150)

Page 18: John Ruskin, *Cascade de la Folie, Chamonix*, 1849, pencil, watercolour and bodycolour, 46.1 x 37.3 cm, Birmingham Museums Trust (1905P2)

Page 20: John Ruskin and John Hobbs, *Mer de Glace, Chamonix*, 1849, daguerreotype, half-plate, 12.1 x 16.2 cm, Whitehouse Collection, The Ruskin, Lancaster University (1996D0075)

Page 21: John Ruskin, *Mer de Glace from the Montanvers Hotel above Chamonix*, 1849, pencil, ink, watercolour, 32.2 x 49.3 cm, King's College, Cambridge; image courtesy of Professor David Hill

Page 22: John Ruskin, part of Aiguille Structure, Plate 29, *Modern Painters IV*, (1856), engraved by J.C. Armytage

Page 25: John Ruskin, *Mer de Glace, Chamonix*, 1874, watercolour, 27.3 x 50 cm, Whitehouse Collection, The Ruskin, Lancaster University (1996P1206)

Page 26 (left): Plants of *Saxifraga cuneifolia* at Brantwood; image by David S. Ingram

Page 26 (right): John Ruskin, *'Francesca dispersa' and 'Francesca terrestris'*, 1877, pencil and brown ink, 16.5 x 10 cm, Whitehouse Collection, The Ruskin, Lancaster University (1996P1282)

Page 28: Stepped path leading to the Precipice rock at Brantwood; image by Howard Hull

Page 29 (left): The Daffodil Meadow at Brantwood; image by David S. Ingram

Page 29 (right): The Alpine Bank at Brantwood; image by Howard Hull

Page 31: John Ruskin, *Study of a Larch Bud*, pencil, watercolour and bodycolour, 19 x 11.3 cm, Whitehouse Collection, The Ruskin, Lancaster University (1996P1264)

Page 34: John Ruskin, *Mountain Rock and Alpine Rose*, 1844 or 1849, ink, watercolour and bodycolour, 30 x 41.2 cm, Whitehouse Collection, The Ruskin, Lancaster University (1996P1395)

Page 40: The fully equipped botanist; frontispiece illustration to Bernard Verlot, *Le Guide du Botanist Herborisant*, (J. B. Baillière et Fils, Paris, 1865)

Page 42: Three British vascula from the collection of the Royal Botanic Garden Edinburgh; image by Lynsey Wilson

Page 43: A European vasculum, from Bernard Verlot, *Le Guide du Botanist Herborisant*, (J. B. Baillière et Fils, Paris, 1865)

Page 45: A nineteenth century plant press, from Bernard Verlot, *Le Guide du Botanist Herborisant*, (J. B. Baillière et Fils, Paris, 1865)

Page 47: A well mounted and labelled herbarium specimen, from Bernard Verlot, *Le Guide du Botanist Herborisant*, (J. B. Baillière et Fils, Paris, 1865)

Page 48: Pressed plant (*Ranunculus glacialis*) from Ruskin's *Flora of Chamouni 1844*; image by Michael Pollard

Page 54: John Ruskin, *Star Gentian*, 1849, pencil, ink and watercolour, 22.8 x 18.5 cm, Whitehouse Collection, The Ruskin, Lancaster University (Diary notebook MS 6)

Page 64: John Ruskin, *Gentians, Les Rousses*, 1882, pencil and watercolour, 12.1 x 20.1 cm, Whitehouse Collection, The Ruskin, Lancaster University (1996P1296)

Page 92: John Ruskin, *Mélèze* (*Pine*), 1844, pencil and watercolour, 23.2 x 18.8 cm, Whitehouse Collection, The Ruskin, Lancaster University (Diary notebook MS 4)

Page 104: John Ruskin, *Drawings of parts of a Conifer Branch*, 1844, pencil and ink, 23.2 x 18.8 cm, Whitehouse Collection, The Ruskin, Lancaster University (Diary notebook MS 4)

Page 132: John Ruskin, *Aspen Tree*, date unknown, pencil and watercolour, 27.2 x 23 cm, Whitehouse Collection, The Ruskin, Lancaster University (1996P1125)

Page 134: Part of the title page of Vol. II, 1835 of W. Baxter, *British Phænogamous Botany*, 1834–43; image by Hazel Drummond

Page 135: John Ruskin, *Oxalis* (Wood Sorrel), date unknown, pencil and watercolour, 12.5 x 9 cm, Whitehouse Collection, The Ruskin, Lancaster University (1996P1262)

Page 137: John Ruskin, *Rhododendron Leaves*, c. 1854–58, pencil, watercolour and bodycolour, 13.3 x 25.5 cm, Whitehouse Collection, The Ruskin, Lancaster University (1996P1352)

Page 138: The six volumes of W. Baxter, *British Phænogamous Botany*, as rebound by Ruskin; image by Hazel Drummond

Page 139 (left): An interleaved page from Ruskin's re-ordered Baxter; image by Hazel Drummond

Page 139 (right): A plate showing a typical 'Land Cinquefoil' in Ruskin's re-ordered Baxter; image by Hazel Drummond

Page 141: Brantwood from the eastern shore of Coniston Water; image by Nina Claridge

Page 142: John Ruskin, *A single petal from a Rhododendron flower at Brantwood*, pencil, 14 x 10.5 cm, Whitehouse Collection, The Ruskin, Lancaster University (1996P1457)

Page 143: John Ruskin, *A dissected inflorescence of Common Self-heal (Prunella vulgaris)*, date unknown, pencil and watercolour, 15.5 x 10.5 cm, Whitehouse Collection, The Ruskin, Lancaster University (1996P1278)

Page 144: The Professor's Garden, Brantwood; image by David S. Ingram

Page 145: The beck, Brantwood; image by David S. Ingram

Page 146: John Ruskin, *Study of Roses, from the dress of Spring in Botticelli's* Primavera, 1874, pencil, ink, watercolour and bodycolour, 16.2 x 24.2 cm, Whitehouse Collection, The Ruskin, Lancaster University (1996P1168)

Page 149: The initial page of Ruskin's index to the re-ordered and rebound volumes of J. E. Smith & J. Sowerby, *English Botany*, 1832–40, usually known as 'Sowerby'; image by Hazel Drummond

Endpapers: Fold-out image from Robert Burford (1791–1861), *Description of a view of Mont Blanc, the valley of Chamounix, and the surrounding mountains: now exhibiting at the Panorama, Leicester Square; painted by the proprietor, Robert Burford, from drawings taken by himself, in 1835.* London, 1837. Printed by T. Brettell

SELECTED BIBLIOGRAPHY AND FURTHER READING

POWO (2019). Plants of the World Online. Facilitated by the Royal Botanic Gardens, Kew. Published on the Internet; http://www.plantsoftheworldonline.org/, retrieved 14th September 2021

Barnes, Janet, 'The Mineral Collection of John Ruskin in the Ruskin Gallery, Sheffield', *UK Journal of Mines & Minerals*, no. 11, Spring/Summer 1992, pp. 26–29

Blamey, Marjorie and Christopher Grey-Wilson, *Wild Flowers of Britain & Northern Europe*, Cassell, London 2003

Brown, David Blayney, *Turner in the Alps 1802*, Tate Gallery, London, 1998

Cook, Edward T. and Alexander Wedderburn (eds.), *The Works of John Ruskin*, (39 vols.), George Allen, London, 1903–1912

Drouet, Serge and Phillipe Hervé Leloupe, *Guides Géologiques: Vallée de Chamonix—Massif du Mont-Blanc*, Omniscience, Montreuil, and BRGM Éditions, Orléans, 2015

Evans, Joan and John Howard Whitehouse (eds.), *The Diaries of John Ruskin*, (3 vols.), Clarendon Press, Oxford, 1956–59

Goepfort, Yvette, *Chamonix: Mont Blanc*, Nouvelles Editions AIO, La Trinité [Alpes-Maritimes], 2010

Grey-Wilson, Christopher and Marjorie Blamey, *Alpine Flowers of Britain and Europe*, 2nd edition, Harper Collins Publishers, London, 1995

Hayman, John, *John Ruskin and Switzerland*, Wilfred Laurier University Press, Waterloo, Ontario, 1990

Hélard, André, *John Ruskin et les Cathédrales de la Terre*, Éditions Guérin, Chamonix, 2005

Hill, David, *Turner in the Alps: The journey through France & Switzerland in 1802*, George Philip, London, 1992

Hill, David, 'Ruskin Drawings at King's College, Cambridge: #4 The Mer de Glace from the Montanvers Hotel above Chamonix', *Sublime Sites*, https://sublimesites.co/2018/12/06/ruskin-drawings-at-kings-college-cambridge-4-the-mer-de-glace-from-the-montanvers-hotel-above-chamonix (retrieved April 2019)

Hoppe, Ansgar, *A Field Guide to the Flowers of the Alps*, Pelagic Publishing, Exeter, 2013

Ikin, Caroline, *Reading Ruskin in the garden: the designed landscape at Brantwood (1871–1900)*, a thesis submitted in partial fulfilment of the requirements of Manchester Metropolitan University for the degree of Doctor of Philosophy, Manchester, 2019

Ingram, David [S.] and Stephen Wildman, *Ruskin's Flora: The Botanical Drawings of John Ruskin*, Lancaster University, Ruskin Library and Research Centre, Lancaster, 2011

Ingram, David [S.], *The Gardens at Brantwood: evolution of John Ruskin's Lakeland paradise*, Pallas Athene, London, 2014

Ingram, David [S.], *Ruskin's Botanical Books: Re-ordered and Annotated Editions of* Baxter *and* Sowerby, Guild of St George, York, 2016

Jacobson, Ken and Jenny, *Carrying Off the Palaces: John Ruskin's Lost Daguerreotypes*, Quaritch, London, 2015

Jermyn, Jim, *Alpine Plants of Europe: a gardener's guide*, Timber Press, Portland, 2005

Mathieu, Jon, *The Alps: An Environmental History*, Polity Press, Cambridge, 2019

Ménard, Thierry, *Grand guide de la flore des Alpes*, Éditions Sud Ouest, Bordeaux, 2007

Nagy, Laszlo, Georg Grabherr, Christian Körner, Desmond B. A. Thompson (eds.), *Alpine Biodiversity in Europe*, Springer, Berlin , 2007

Overal, Bernard, *Guide de la flore des Alpes Françaises*, 2nd edition, Naturalia Publications, Turriers [Alpes-de-Haute-Provence], 2016

Payot, Venance, *Herbier des Alpes*, Les Éditions de l'Amateur, Paris, 2006

Sdegno, Emma, 'The Alps', in Francis O'Gorman (ed.), *The Cambridge Companion to John Ruskin*, Cambridge University Press, Cambridge, 2015

Tilson, Jean-Marc and Bruno de Foucault, *Flora Gallica: Flore de France*, Biotope Editions, Mèze [Hérault], and Société Botanique de France, Paris, 2014

Tutin, Thomas G., Vernon H. Heywood, and others, *Flora Europaea*, (5 vols.), Cambridge University Press, Cambridge 1964–1980; 2nd edition 1993–

Warrell, Ian, *Through Switzerland with Turner: Ruskin's First Selection from the Turner Bequest*, Tate Gallery, London, 1995

Wildman, Stephen, *Mountain Glory: Ruskin's 'Modern Painters' and the Swiss Alps*, Lancaster University, Ruskin Library and Research Centre, Lancaster, 2010

Wildman, Stephen, *"Beautiful Effects": Ruskin's Daguerreotypes of Switzerland*, Lancaster University, Ruskin Library and Research Centre, Lancaster, 2012

ACKNOWLEDGEMENTS

The idea of producing an edition of the *Flora of Chamouni 1844*, fascinating in its own right and containing one of the most significant amounts of Ruskin's unpublished manuscript material, grew out of the exhibition *Ruskin's Flora: The Botanical Drawings of John Ruskin*, held in 2011 at the Ruskin Library and Research Centre (now The Ruskin), Lancaster University, with an accompanying conference at the Lancaster Environment Centre. Those involved then have also contributed hugely to this project, especially Rebecca Patterson, Diane Tyler and the late Jen Shepherd, followed by their successors at the Ruskin, Professor Sandra Kemp and Dr Christopher Donaldson. Help in sourcing additional images for Stephen Wildman has been kindly given by the late Dr James Dearden, Professor André Hélard, Rennes, and Professor David Hill, University of Leeds.

New photography of exceptional quality has been undertaken by Michael Pollard,on the helpful recommendation of Carol Burrows, Rylands Library, University of Manchester. This and publication of the *Flora* have been made possible by a generous grant from the Finnis Scott Foundation, augmented by support from the Dorothy Howe Gift.

Alexander Fyjis-Walker's professional design and editorial skills are evident throughout the book and we thank him warmly for his contribution.

David S. Ingram also wishes to thank the following. For privileged access to the scientific staff and facilities of the Royal Botanic Garden Edinburgh (RBGE): the Trustees and Simon Milne MBE, Regius Keeper. For invaluable advice in the identification of specimens in John Ruskin's *Flora of Chamouni 1844*: Dr David Harris, Director of the Herbarium, RBGE and his team; also, Dr Alan Elliott, Dr David Long, Dr Philip Lusby MBE, and Dr Henry Noltie (present or former Scientific staff of the RBGE); and Jim Jermyn, Property Manager/Head Gardener, Branklyn Garden, National Trust for Scotland; (it must be noted, however, that any errors in the identification of specimens are DSI's sole responsibility). For equally invaluable library assistance: Lorna Mitchell, Head of Library, Archives and Publications, RBGE and her library team, especially Graham Hardy, Serials Librarian and Lynsey Wilson, Photographer. For many fascinating discussions about Ruskin's garden-making: Howard Hull, Director of Brantwood and Dr Caroline Ikin, a Curator of Designed Landscapes, National Trust. For innumerable botanical discussions and/or academic support: John Parker VMH, FLS, Professor (now emeritus) of Plant Cytogenetics and former Director, University of Cambridge Botanic Garden; Nigel Paul, Professor (now emeritus) in Plant Biology, Lancaster (University) Environment Centre; and Steven Yearley, FRSE, Director of the Institute for Advanced Studies in the Humanities and Professor of the Sociology of Scientific Knowledge, Science, Technology and Innovation Studies, the University of Edinburgh. For generous and invaluable editorial assistance: A. I., North Berwick.

D.S.I.

S.G.W.

INDEX OF PLANT SPECIES IN RUSKIN'S *FLORA OF CHAMOUNI 1844*

The key species names (printed in blue) in this index are those given to specimens by Ruskin, spelt as he spelt them, in his *Flora of Chamouni 1844*. In those cases where Ruskin did not give a name to a specimen, the name used is that given by me (DSI).

The page numbers given against the key species names are to the pages in the present volume where the specimens are described and discussed. These pages, or pages adjacent to them, provide cross references to the relevant specimen pages in the *Flora of Chamouni 1844*.

Acinos alpinus see *Thymus alpinus*
Alchemilla acutiloba see *Alchemilla vulgaris*
Alchemilla vulgaris 74
Androsace alpina 123
Androsace carnea 100
Androsace laggeri see *Androsace carnea*
Antennaria dioica see *Gnaphalium dioicum*
Astrantia minor 125
Beckwithia glacialis see *Ranunculus glacialis*
Bistorta officinalis 82
Bistorta vivipara 82
Cardamine crassifolia see *Cardamine pratensis*
Cardamine palustris see *Cardamine pratensis*
Cardamine pratensis (possibly a variant) 95
Carex sp. possibly *Carex nigra* 81
Cherleria laricifolia 111
Clinopodium alpinum see *Thymus alpinus*
Crepis aurea 96
Draba aizoides 58
Epilobium fleischeri 120
Epilobium montanum 128
Erigeron uniflorus 117
Eriophorum scheuzeri 82, 83
Gagea bohemica see *Ornithogalum fistulosum*
Gagea fragifera see *Ornithogalum fistulosum*
Gentiana alpina 61
Gentiana clusii see *Gentiana alpina*
Geum montanum 68
Globularia cordifolia 61, 62, 63
Gnaphalium dioicum 59, 60, 77, 113, 115
Gymnadenia conopsea see *Orchis nigra*
Gymnadenia rhellicani see *Orchis nigra*
Homogyne alpina see *Tussilago alpina*
Hornungia alpina or, depending on pH of the substrate, *H. alpina* subsp. *brevicaulis* 116
Juncus campestris see *Luzula campestris*
Kalmia procumbens 69
Ligusticum mutellinoides see *Neogaya simplex*
Linaria alpina 121
Loiseleuria procumbens see *Kalmia procumbens*
Lonicera alpigena see *Lonicera nigra*
Lonicera nigra (possibly) 97
Lonicera xylosteum see *Lonicera nigra*
Luzula campestris 84
Luzula lutea 117
Maianthemum bifolium see *Myanthemum bifolium*
Medicago lupulina 130
Menyanthes trifoliata 126

Minuartia laricifolia see *Cherleria laricifolia*
Moehringia sp. (possibly *Moehringia muscosa*) 76, 77
Myanthemum bifolium 57
Neogaya simplex 115
Orchis nigra 71, 72, 98
Ornithogalum fistulosum 67
Orobanche spp. 63
Pinguicula alpina 58, 86, 105, 106
Polygala alpestris see *Polygala alpina*
Polygala alpina 91
Polygala chamaebuxus 110
Polytrichastrum alpinum 87
Polytrichum alpinum see *Polytrichastrum alpinum*
Polytrichum commune 86
Potentilla aurea 73
Potentilla crantzii see *Potentilla aurea*
Primula farinosa 81
Primula integrifolia 71
Ranunculus glacialis 115, 131
Ranunculus kuepferi see *Ranunculus pyrenaeus*
Ranunculus pyrenaeus 70
Rorippa palustris see *Cardamine pratensis*
Saxifraga aizoon see *Saxifraga cotyledon*
Saxifraga aspera 89
Saxifraga bryoides 116
Saxifraga cotyledon 88
Saxifraga cuneifolia 78
Saxifraga exarata see *Saxifraga muscoides*
Saxifraga moschata see *Saxifraga muscoides*
Saxifraga muscoides 90, 123
Saxifraga oppositifolia 122
Saxifraga paniculata see *Saxifraga cotyledon*
Saxifraga rotundifolia 80
Scleranthus perennis 110
Soldanella alpina 59, 70
Thalictrum aquilegifolium 102
Thalictrum aquilegiifolium see *Thalictrum aquilegifolium*
Thymus alpinus 129
Tussilago alpina 101
Tussilago farfara see *Tussilago alpina*
Veronica fruticans 125
Veronica urticifolia 112

GENERAL INDEX

Acland, Henry 16
Allen, George 146
Alpine botany
  Zonation and life forms 30–33
  Naming alpine plants 35–39
Baveno 127
Baxter, W., *British Phænogamous Botany* (1834–43) 7, 133, 137, 147
Beamish, Sally 29
Blackburn, Jemima 17
Blunt, Wilfrid 148
Bonneville 57
Brantwood 27, 135, 141–45, 147
Brown, John 23
Brown, Rev. W. L. 13
Burgess, Arthur 146
Chamonix and environs
  Argentière, 51, 65, 105
  Arveyron, source, 76, 86, 124, 151
  Benit, lake and mountain 58–59, 61
  Bérard, Col 59, 83, 105, 151
  Blaitière, Glacier 100
  Bois, Glacier des 88–89, 151
  Bossons, Glacier des 20, 22, 103, 109, 118, 120–21, 151
  Brévent, Le 71, 108–10, 112–13, 118, 129, 151
  Buet 65, 88, 105
  Chapeau, Le 86, 88, 106, 118
  Charmoz, Aiguilles 99–100, 102, 151
  Côte, Montagne de La 115–17
  Entragues, Val 59, 105–06
  Flégère 59, 66, 71, 151
  Goûter, Dôme du 114
  Hotel de l'Union 19
  Houches, Les 125
  Mer de Glace 11, 20, 22, 151
  Midi, Aiguille du 118, 120, 122
  Mont Blanc 11, 14, 24, 66, 69, 118, 122, 127, 151
  Montenvers 11, 17, 19–20, 94, 99–100, 124–25, 151
  Pèlerins, Cascade and Glacier 19, 75–76, 78, 87, 93–98, 103, 105, 108, 123, 131, 151
  Plan, Aiguille du 118
  Prieuré 76, 102, 106
  Reposoir, Mont 53, 58, 63, 73, 99
  Servoz 127
  Taconnaz, Glacier 109, 115, 124, 151
  Tapia, La 75, 108, 118, 151
  Tête Noire 79–80, 83, 125, 128, 151
  Valorsine 130
Clayton, Edward 13
Cluse 57
Collingwood, W. G. 142
Cook, E. T. 7
Couttet, Joseph-Marie 14, 24, 36, 41, 52–53, 65, 81, 83–85, 88, 94, 99, 108–09, 118, 122, 124–25, 127, 131
Crawley, Frederick 21
Curtis, W., *Botanical Magazine* (1795) 133, 146
Daguerreotypes 19, 21, 23
Devouassoud, Michel 65
Dillennius, Johan Jacob 41
Dumas, Alexandre 16
Eisenkrämer, family 19
Florence 15–16
Forbes, James 14
Geneva 15, 55

Giessbach 24
Great St. Bernard Pass 11
Hobbs, John, known as George 19, 41, 65, 85, 94, 108–09, 124
Humboldt, Alexander von 30
Les Rousses 55
Linnæus, Carolus (Carl von Linné) 35, 107, 143
Lucca 16
Lucerne 23
Matterhorn 19
Millais, John Everett 20
Moore, Rev. Daniel 20
Mornex 8, 23
Nant d'Arpanaz 63
Newton, Charles 20
Norton, Charles Eliot 23–24, 145
Oeder, Georg Christian 36, 140
Oxford, Christ Church 12
Payot, Pierre Joseph 16
Payot, Venance 8, 15–16, 36, 46, 49
Pisa 16
Plant collecting in the nineteenth century
  Vascula 41–43
  Recording 43
  Ordering and pressing specimens 44–46
  Mounting and treating specimens 46–47
  Ruskin's motives and criteria in making his alpine collection 49–53
Richardson, Mary 11, 16
Richardson, William 15–16
Richmond, George 16
Rousseau, Jean-Jacques 36
Ruskin, Effie 20
Ruskin, John
  TOURS
  1833, 1835 11
  1840–41 12
  1842 12–13
  1844 14
  1846 16
  1849 17, 19–20
  1854 20–21
Ruskin, John, TOURS *cont.*
  1856 21–22
  1858, 1860 23
  1861 23
  1862–63 23
  1866 24
  1874 24
  1888 24
  WORKS
  A Tour through France (1835) 12
  Facts and considerations on the strata of Mont Blanc (1834) 11
  Modern Painters (1843–60) 8, 13–14, 16, 21–23, 25, 105, 136, 147
  On the Stratified Alps of Savoy (1863) 9
  Poems 11, 14
  Præterita (1885–89) 12–14, 16, 50, 134
  Proserpina (1875–86) 7, 133, 135, 137, 140, 145–48, 150
  Unto This Last 23, 25
Ruskin, John James 9, 11, 14, 24
Ruskin, Margaret 9, 11, 63, 133, 138
St. Gothard Pass 13
St. Martin 55
Saussure, Horace Bénédict de 11, 99, 130
Severn, Arthur 24
Severn, Joan 24, 29
Simon, Mrs John 24
Simplon 11, 14, 121
Smith, J. E. and Sowerby, J., *English Botany* (1832–40) 7, 148–49
Splügen Pass 11
Stearn, William 148
Stillman, William James 23
Trevelyan, Pauline 24, 137
Turner, J. M. W. 13, 136
Venice 16–17, 20, 24
Verlot, Bernard 44, 46
Verona 17
Waterloo 11
Watson, H. C. 148–49
Welsh, Jane (Mrs Thomas Carlyle) 136–38
Whitehouse, John Howard 7
Withering, William 41, 44, 46
Zermatt 14, 127

FLORA OF CHAMONIX by John Ruskin,
edited by David S. Ingram and Stephen Wildman

Volume I: Facsimile ISBN 978 1 84368 231 8
Volume II: Commentary ISBN 978 1 84368 232 5
Set: ISBN 978 1 84368 233 2

First published 2023 by Pallas Athene (Publishers) Ltd
2 Birch Close, Hargrave Park, London N19 5XD

Book design and layout by Alexander Fyjis-Walker

Printed in England

A Description of A View *of* Mont Blanc,

1. Aiguille Verte.
2. Aiguille du Dru.
3. Glacier du Nant Blanc.
4. Le Chapeau.
5. Mer de Glace.
6. Glacier des Bois.
7. Aiguille de Bochard.
8. Rochers des Mottets.
9. Aiguille du Moine.
10. Petite Jorasse.
11. Glacier de Léchaud.
12. Grand Jorasse.
13. Les Periades.
14. Le Tacul.
15. Le Jardin.
16. Montanvert.
17. Sentier de la Filia.
18. Source de L'Arveiron.
19. Hameau des Bois.
20. Sentier du Montanvert.
21. Hameau des Mouilles.
22. Route de Martigny.
23. Hameau des Prés.
24. L'Arve.
25. La Prieuré de Chamounix.
26. Chalets sur le Rocher.
27. Valley of Chamounix.
28. Hameau des Pélerins.
29. Cascade des Pélerins.
30. Pont Piralotta.
31. Hameau des Buissons.
32. Mont Mallet.
33. Le Géant.
34. Aiguille des Charmoz.
35. Glacier de Crepon.
36. Aiguille de la Blaitière.

65. Cabane de la Croix de Flégère.
66. Aiguilles Rouges.
67. Chaine Septentrionale du Valais.
68. La Genousi.
69. Rochers de Croix de Fer.
70. L' Hospice du Col de Balme.
71. Chalets de Charamillan.